BASICS OF HOSPITALITY ENGINEERING

BASICS OF HOSPITALITY ENGINEERING

A Practical Guide for Future Hotel Managers

Eric Bouf

ÉK::LÏT Hospitality Collection

FSC
www.fsc.org
MIXTE
Papier issu
de sources
responsables
Paper from
responsible sources
FSC® C105338

Édition : BoD · Books on Demand, 31 avenue Saint-Rémy, 57600 Forbach, bod@bod.fr

Impression : Libri Plureos GmbH, Friedensallee 273, 22763 Hambourg (Allemagne)
Impression à la demande
ISBN : 978-2-3228-5841-5
Dépôt légal : Juin/2026

“This book is dedicated to the students who will one day carry the responsibility of managing a hotel, not only through service, but through judgment, vigilance, and care.“

Table of Contents

ACKNOWLEDGEMENTS

This book was created to support future hospitality managers in understanding a part of hotel operations that is often invisible until something goes wrong: engineering.

I would like to acknowledge the many hotel engineers, technicians, operations leaders, General Managers, asset managers, owners, contractors, and colleagues whose work has shaped my understanding of hospitality as an ecosystem. Their daily commitment protects the guest experience in ways that are rarely noticed, yet deeply essential.

I also wish to thank the students whose questions continue to remind me that good teaching is not about simplifying reality, but about making complex responsibilities accessible. Their curiosity, doubts, and practical intelligence are at the heart of this book.

Finally, this work reflects a conviction developed through years of hotel operations and teaching: future hospitality leaders do not need to become engineers, but they must understand engineering well enough to ask the right questions, defend the right budgets, and protect the guest, the team, the asset, and the brand.

ABOUT THE AUTHOR

Hospitality professional, educator, and consultant with more than thirty years of international experience in hotel management across Europe, Asia, the Americas, Dubai, and North Africa.

After holding senior operational and leadership roles in the hospitality industry, I now teach hospitality management to postgraduate students in France. My teaching focuses on helping future managers connect operational reality, leadership responsibility, financial awareness, and guest experience.

I root my academic work in a practical conviction: hotel managers do not need to become technical experts in every field, but they must understand enough to ask the right questions, challenge assumptions, support their teams, and make informed decisions.

Through this book, I bring hospitality engineering into a managerial perspective. The objective is not to train engineers, but to prepare future General Managers, Operations Directors, Asset Managers, owners' representatives, and hospitality leaders to understand engineering as a strategic function that protects people, revenue, continuity, reputation, and long-term asset value.

The teaching approach combines operational realism, executive posture, and applied case-based learning. It invites students to move beyond technical vocabulary and to understand engineering as a discipline of prevention, coordination, accountability, and decision-making.

INTRODUCTION — WHY ENGINEERING MATTERS IN HOSPITALITY

Hospitality is often described through what guests see and feel: the welcome at the entrance, the comfort of the room, the quality of the restaurant, the confidence of the staff, the quiet efficiency of service. Yet behind these visible moments stands another reality. A hotel is also a building that never sleeps. It is a complex technical organism where air, water, electricity, heat, light, safety systems, digital infrastructure, waste flows, elevators, kitchens, laundries, equipment rooms, guest rooms, and public areas must work together every hour of every day.

When everything functions properly, engineering remains invisible. Guests do not usually praise the hot water system when the shower is perfect. They do not thank the electrical distribution network when the lights work. They do not think about the air-handling units when they sleep well. They rarely notice the preventive maintenance program that kept the elevator operating, the emergency power testing that protected life-safety systems, or the engineering reporting that warned management before a breakdown became a crisis.

But when engineering fails, it becomes immediately visible. A room that is too hot, a guest corridor without lighting, a broken elevator, a plumbing leak, a fire alarm malfunction, a kitchen equipment failure, a power outage, poor air quality, a safety inspection issue, or an uncontrolled renovation delay can quickly transform a technical matter into a guest-experience problem, a financial exposure, a staff-stress issue, a compliance concern, and a reputational risk. In hospitality, technical failures rarely stay technical for long.

This is why engineering matters to future hotel managers. The purpose of this book is not to turn hospitality students into engineers. It is to help them become credible, informed, and responsible managers who understand enough to ask the right questions, challenge superficial answers, prioritize risks, defend budgets, communicate with owners, and support their engineering teams with clarity.

Engineering as the Invisible Foundation of Guest Experience

A guest experience is never only emotional. It is also physical. Comfort is produced by infrastructure. A good night's sleep depends on temperature, air quality, sound insulation, lighting, plumbing, bedding, and maintenance discipline. A smooth check-in depends not only on reception skills, but also on electricity, network availability, access control systems, elevators, and room readiness. A successful event depends on power, ventilation, kitchen equipment, fire safety, water supply, audiovisual reliability, and fast coordination when something goes wrong.

The guest rarely separates these elements. If the air conditioning is noisy, the room is "bad." If the shower has poor pressure, the hotel is "not worth the price." If the elevator fails during peak arrival time, the property feels poorly managed. If a fire alarm creates confusion, the issue becomes trust. Infrastructure therefore influences the emotional memory of the stay. It does not always create delight, but its failure creates disappointment immediately.

For this reason, engineering should not be seen as a department that simply repairs defects. It is a department that protects the conditions under which service can exist. Front Office can only welcome properly if rooms and systems are functional. Housekeeping can only deliver quality if equipment, water, lighting, ventilation, and room assets are reliable. Food and Beverage depends on kitchens, extraction, refrigeration, power, fire protection, and safe back-of-house systems. Security depends on alarms, access systems, emergency lighting, and communication infrastructure. Finance depends on energy control, maintenance planning, asset protection, and capital discipline. Engineering touches every department indirectly because every department depends on the building.

From Maintenance Thinking to Management Thinking

Many people reduce engineering to maintenance. This is understandable, because maintenance is the most visible expression of engineering work. A guest calls because something is broken; a technician arrives; the issue is fixed. Yet this view is incomplete. A strong engineering function is not primarily reactive. Its real value lies in prevention, anticipation, planning, documentation, risk reduction, and asset lifecycle management.

For hotel leaders, the important question is not only "Can this be repaired?" but also "Why did it fail?" "Was it predictable?" "Was preventive maintenance funded?" "How many rooms are exposed?" "What is the risk if we wait?" "Is this an isolated defect or the sign of a larger system problem?" "What will this cost if we postpone the decision?" "Is the issue affecting guest satisfaction, compliance, energy efficiency, safety, or asset value?" These are managerial questions, not technical questions.

A future General Manager does not need to know how to replace a pump, recalibrate a chiller, test a sprinkler line, or diagnose an electrical panel. However, the same General Manager must understand the consequences of delayed maintenance, weak reporting, poor documentation, underfunded preventive programs, and unclear accountability. Leadership begins where technical detail becomes business impact.

Engineering as Risk Management

Engineering is one of the major risk-management functions of a hotel. It protects operational continuity by keeping the property open and functional. It protects guest comfort by maintaining systems that guests experience directly, even when they do not see them. It protects safety by ensuring that alarms, emergency lighting, fire doors, suppression systems, evacuation routes, and critical equipment are inspected and maintained. It protects financial performance by controlling energy use, reducing emergency repair costs, extending asset life, and avoiding unplanned room closures. It protects reputation by preventing technical failures from becoming public complaints or media narratives.

The manager's responsibility is therefore to understand risk exposure. A small leak is not only a plumbing defect if it can close several rooms, damage ceilings, disturb guests, create mold, trigger compensation, affect online reviews, and require capital repair. A delayed fire-door inspection is not only a checklist issue if it weakens evacuation logic and exposes the hotel to legal and insurance consequences. An old HVAC system is not only an engineering inconvenience if it increases energy costs, produces comfort complaints, and creates the possibility of large-scale failure during peak season.

Risk in hospitality is often cumulative. Problems become dangerous when they are normalized. A minor defect is postponed, then repeated, then patched, then hidden inside the operating budget, until it becomes a major capital problem or a crisis. Good engineering management interrupts this pattern. It makes weak signals visible before they become emergencies.

The Financial Dimension of Engineering

Engineering is sometimes perceived as a cost center. This perception is dangerous because it encourages underinvestment. In reality, engineering protects revenue and margin. It keeps rooms sellable. It reduces guest compensation. It prevents avoidable breakdowns. It supports energy efficiency. It protects asset value. It helps managers plan capital expenditure instead of being trapped in emergency operating expense.

The distinction between CAPEX and OPEX is central to hotel engineering. Some decisions require capital investment because the asset has reached the end of its useful life or because legal, safety, or brand requirements demand a structural response. Other decisions belong to operating expenses because they concern regular maintenance, routine service, or short-term repairs. The difficulty for managers is that postponed CAPEX often returns as exploding OPEX: repeated repairs, emergency call-outs, energy waste, guest compensation, staff overtime, and operational disruption.

Future managers must learn to defend engineering budgets with business logic. Owners and finance teams may not respond to technical language alone. They need to understand risk, return, timing, cash impact, depreciation logic, guest satisfaction consequences, and asset value. A strong hotel manager translates engineering needs into financial consequences.

Safety, Accountability, and Leadership

Fire, life safety, and regulatory compliance reveal the deepest responsibility of hospitality leadership. A hotel is a place where guests sleep in an unfamiliar environment. They trust the property with their safety, often while they are at their most vulnerable. Fire detection, suppression, emergency lighting, evacuation routes, compartmentation, staff training, maintenance logs, and compliance certificates are not administrative details. They are part of the moral and legal contract between the hotel and its guests.

Managers can delegate tasks, but they cannot delegate accountability. Engineers test systems. Contractors service equipment. Specialists inspect and certify. Yet the hotel remains responsible for ensuring that the work is planned, funded, documented, reviewed, and acted upon. “I did not know” is not a strong managerial defense. A leader must create the reporting structure that allows important risks to be seen, escalated, and resolved.

This book therefore treats safety not simply as a technical topic, but as a leadership discipline. Safety failure is often preceded by management failure: deferred maintenance, incomplete training, unclear command structure, weak documentation, poor communication, and budget decisions that underestimated risk.

Communication: Speaking Engineering as a Manager

One recurring problem in hotels is that Engineering and Operations do not always speak the same language. Engineering teams speak in systems, equipment, faults, tolerances, inspections, readings, and interventions. Management teams speak in guest impact, risk level, cost, timing, reputation, revenue, and ownership decisions. When the translation is weak, decisions become poor.

This book encourages future managers to develop operational vocabulary. The goal is not jargon. The goal is clarity. Words such as corrective maintenance, preventive maintenance, predictive maintenance, lifecycle cost, deferred maintenance, asset integrity, phasing strategy, emergency power, passive fire protection, and operational continuity allow managers to name problems more accurately. When problems are named accurately, they can be prioritized, budgeted, explained, and solved more effectively.

A good manager does not hide behind technical complexity. A good manager asks for a clear description of the issue, the consequence of inaction, the stakeholders affected, the options available, the recommended decision, and the timing required. A technical incident only becomes useful to management when it has been translated into business impact.

People, Structure, and the Engineering Team

Engineering is also a people-management function. Even the best technical systems require competent, organized, motivated, and accountable teams. Hotels must decide how much expertise to keep internally, what to outsource, how to structure shifts and on-call duties, how to document performance, how to manage subcontractors, and how to select Engineering Managers who can lead rather than simply repair.

The best technician is not automatically the best Engineering Manager. Technical excellence is essential, but leadership requires additional capabilities: communication, planning, prioritization, budgeting, documentation, vendor governance, cross-functional coordination, safety discipline, and executive reporting. Engineering Managers operate at the intersection of people, systems, finance, compliance, and operations.

In many hotels, the local engineering team carries the burden of aging infrastructure, high guest expectations, limited capital, urgent breakdowns, owner pressure, and regulatory obligations. Their work deserves managerial respect. A General Manager who understands engineering can support the team more effectively, challenge it more intelligently, and defend its priorities more credibly.

How This Book Is Organized

The book follows a practical progression. It begins with the fundamentals of the Engineering Department: its purpose, structure, responsibilities, and accountability. It then examines the building infrastructure and technical systems that make a hotel function, including HVAC, plumbing, electricity, emergency power, energy distribution, IoT, waste management, and air quality. The third chapter focuses on fire, life safety, and regulatory constraints as managerial responsibilities. The fourth chapter develops the vocabulary and communication skills needed to translate technical issues into executive decisions. The fifth chapter addresses renovation programs, CAPEX and OPEX logic, hidden costs, phasing, and financial arbitration. The sixth chapter focuses on local engineering team management, recruitment, subcontracting, shift organization, performance, and the role of the Engineering Manager.

Each chapter is written from the perspective of hospitality management. The emphasis is on judgment, responsibility, risk, guest impact, financial logic, and operational coordination. The operational case studies invite readers to think like General Managers, Executive Committee members, asset managers, owners' representatives, and future leaders facing real constraints.

By the end of this book, readers should be able to understand engineering challenges without becoming engineers, make informed decisions related to building systems, manage engineering teams with greater confidence, defend renovation and maintenance budgets, and act as credible interlocutors with owners, insurers, regulators, contractors, and technical partners.

A Final Word Before Beginning

The invisible parts of hospitality often determine the visible quality of service. Engineering may not always appear in guest comments when it succeeds, but it is present in every comfortable room, every safe corridor, every functioning kitchen, every reliable lift, every compliant building system, and every prevented crisis.

For future hotel managers, engineering is not a subject to fear. It is a language to learn, a responsibility to accept, and a discipline to respect. You do not need to fix the machine yourself. But you must understand what the machine protects, what happens when it fails, and what decisions are required before failure reaches the guest.

Chapter 1

Engineering Department Fundamentals

Purpose, Structure & Responsibilities

In a hotel, engineering is often the department guests never see until something breaks. The guest does not normally ask who maintained the air-conditioning system, who checked the emergency lighting, who tested the pumps, who monitored the boiler, or who coordinated the elevator inspection. The guest simply expects the room to be comfortable, the water to be hot, the lights to work, the Wi-Fi to be reliable, and the building to feel safe. When those conditions are met, engineering remains invisible. When they are not met, engineering suddenly becomes one of the most visible departments in the hotel.

This chapter introduces the Engineering Department as a managerial function, not as a purely technical workshop. The purpose is not to transform hospitality students into engineers. It is to help future hotel managers understand enough about engineering to ask the right questions, interpret risks, support technical teams, communicate with owners and insurers, and make better operational decisions. A General Manager does not need to repair a chiller, rewire an electrical panel, or certify a fire system. However, the General Manager must understand the business consequences if these systems fail, are underfunded, poorly maintained, or managed without clear accountability.

The central idea of this chapter is simple: engineering is business continuity. Hotels are 24-hour operations where guest satisfaction, revenue protection, safety, compliance and asset value depend on infrastructure that must perform continuously. Rooms can only be sold if they are functional. Restaurants can only serve if kitchens, refrigeration and ventilation systems operate correctly. Spas, meeting rooms, back-of-house areas, laundries and public spaces all rely on technical systems. Engineering therefore supports every guest-facing department, every revenue stream and every promise made by the brand.

Engineering also protects the owner's asset. Hotels are capital-intensive buildings. The value of the property depends not only on design, location or brand reputation, but also on the condition and reliability of the building systems. Poor maintenance accelerates asset depreciation. Deferred repairs create hidden liabilities. Short-term savings can become long-term capital expenditure. A cracked pipe, an ignored vibration, an overdue inspection or a poorly documented repair may appear minor in isolation, but together they create a pattern of risk. Strong engineering management protects both operations and property value.

For this reason, engineering should not be seen only as a cost center. It does not generate revenue in the same direct way as rooms or food and beverage, but it prevents revenue loss. It keeps rooms sellable, equipment reliable, public areas operational and safety systems compliant. It supports margins through energy efficiency, preventive maintenance and better lifecycle planning. The most profitable engineering decision is often the failure that never occurs, the complaint that never reaches reception, the refund that never needs to be offered, and the emergency capital expenditure that never becomes necessary.

The chapter is organized around five managerial questions. First, why does a hotel need an Engineering Department? Second, what are its core missions? Third, how does engineering support other departments? Fourth, how should engineering teams be structured, including the balance between generalists, specialists, internal employees and contractors? Finally, what is the role of hierarchy and accountability, especially for the General Manager? These questions form the foundation for later chapters on infrastructure, fire and life safety, technical vocabulary, renovations, CAPEX and local engineering team management.

Engineering as the Invisible Backbone of Hospitality

The hospitality industry is built on visible experiences: a warm welcome, a clean room, a well-presented meal, a smooth check-in, a calm spa treatment, a memorable event. Behind those visible experiences sits a technical infrastructure that must be planned, maintained and monitored. Engineering is the department responsible for this infrastructure. Its work is often hidden behind walls, ceilings, plant rooms, service corridors, roofs and technical shafts, but its impact is felt in every part of the hotel.

A luxury bedroom is not comfortable because of decoration alone. It is comfortable because temperature, ventilation, water pressure, lighting, acoustic conditions, electrical supply, drainage and safety systems are functioning correctly. A restaurant kitchen does not produce service only because chefs are skilled. It also depends on gas or electrical supply, extraction, refrigeration, dishwashing equipment, grease management, drainage, lighting, water temperature and equipment maintenance. A conference event does not run smoothly only because the sales and banqueting teams are organized. It requires power, air-conditioning, lighting, audiovisual support, fire safety compliance and functioning restrooms.

Engineering therefore gives practical reality to the hotel's commercial promise. If the brand promises comfort, engineering helps deliver comfort. If the hotel promises safety, engineering helps maintain safety systems. If the hotel promises wellness, engineering protects air quality, water quality, temperature control and spa equipment. If the property promises sustainability, engineering becomes central to energy monitoring, water saving, waste systems, preventive maintenance and long-term efficiency.

The best engineering is often invisible because guests notice outcomes rather than systems. They notice that the room is too hot, but not that the chiller is correctly maintained. They notice that the shower is cold, but not that the boiler is functioning properly. They notice that the elevator is out of service, but not the inspection program that usually prevents breakdowns. This visibility gap creates a leadership challenge: because engineering is not always visible, it is sometimes undervalued until failure occurs. Effective managers must learn to value prevention before crisis makes the need obvious.

Why Hotels Need an Engineering Department

Hotels need an Engineering Department because hotel operations cannot stop. Unlike many businesses, a hotel is occupied day and night. Guests sleep, shower, work, dine, move through the building, use elevators, adjust room temperature, charge devices, connect to Wi-Fi, attend events and expect support at any hour. Technical systems must therefore operate continuously, and when they fail, the impact is immediate.

The first reason hotels need engineering is business continuity. A hotel cannot sell rooms that are out of order. It cannot host events if key spaces are unsafe or uncomfortable. It cannot protect revenue if technical failures generate refunds, cancellations or reputational damage. Business continuity means ensuring that systems are sufficiently reliable to support uninterrupted operations. Engineering achieves this through preventive maintenance, monitoring, repairs, emergency response and technical planning.

The second reason is guest safety and comfort. Engineering contributes to safe exits, functioning alarms, emergency lighting, water quality, electrical safety, air-conditioning, ventilation, hot water, elevators and other systems that affect physical safety and comfort. Safety and comfort are not optional luxuries. They are basic conditions for operating a hotel responsibly. A guest may forgive a small service delay, but serious safety failures can create legal, financial and ethical consequences.

The third reason is asset protection. A hotel building is an expensive asset with a long lifecycle. Technical systems must be maintained, replaced and upgraded at the right time. When maintenance is postponed, equipment becomes less efficient, breakdowns become more frequent, and replacement costs can become urgent. Engineering protects the asset by ensuring that the building does not deteriorate silently. This requires documentation, inspections, planned maintenance, budgeting and communication with ownership.

The fourth reason is cost control. Engineering influences utility costs, repair costs, replacement costs, contractor costs and the efficiency of building operations. Energy is often one of the major controllable expenses in a hotel. Poorly maintained equipment consumes more energy and fails more often. Reactive repairs usually cost more than planned maintenance because they happen under pressure, outside normal hours, with operational disruption and limited negotiation power. Preventive maintenance is therefore a financial discipline, not only a technical habit.

The fifth reason is compliance. Hotels must comply with safety codes, maintenance obligations, insurance requirements, local regulations and inspection schedules. Even when specialists perform the technical work, the hotel remains responsible for ensuring that obligations are met and properly documented. Compliance is not simply a matter of avoiding fines. It is part of the hotel's duty of care toward guests, employees and the public.

The Cost of Failure: From Technical Problem to Business Crisis

One of the most important lessons for future managers is that engineering failures rarely remain purely technical. A small failure can quickly become a guest complaint, then an operational disruption, then a financial loss, then a reputational issue, and sometimes a legal or insurance matter. This is why engineering must be understood as risk management.

Consider a broken elevator. Technically, it may be a mechanical or control system issue. Operationally, it affects guest movement, housekeeping logistics, luggage delivery, room service, staff efficiency and accessibility. Financially, it may result in compensation or lost bookings if guests complain. Reputationally, it may generate negative reviews. Legally, if the elevator was overdue for inspection or maintenance, the hotel may face liability. The original problem is technical, but the consequences are managerial.

Consider an air-conditioning failure affecting forty rooms at 2 a.m. during summer. At first, it is a climate control problem. Within minutes, it becomes a front office crisis. Guests call reception. Some ask to move rooms. Some request refunds. Some post online. Night staff must manage complaints without full decision-making authority. Housekeeping may need to prepare replacement rooms. Revenue management loses inventory. The General Manager may face pressure the next morning from guests, owners and brand representatives. The engineering failure has crossed departmental boundaries.

The same applies to water leaks, electrical outages, fire alarm malfunctions, boiler failures, refrigeration breakdowns or emergency system defects. Each of these can create a cascade effect. Effective engineering leadership aims to interrupt this cascade before it reaches the guest. That is why the key sentence for this chapter is: engineering does not only fix problems; it prevents disasters.

Engineering Compared with Other Hotel Departments

Every hotel department has a primary focus. Front Office manages guest contact, arrival, departure, billing, information and complaint handling. Housekeeping manages cleanliness, room readiness and the physical presentation of guestrooms and public areas. Food and Beverage generates revenue through restaurants, bars, banqueting and room service. Sales and Marketing create demand. Finance monitors performance, controls costs and supports business planning. Human Resources supports staffing, compliance, training and employee relations.

Engineering is different because it supports the operating conditions of all these departments. It does not usually own the guest relationship, the room cleaning process or the dining experience, but without engineering these processes cannot function reliably. Engineering supports Front Office by making rooms sellable and responding to technical complaints. It supports Housekeeping by maintaining equipment, lighting, plumbing and room conditions. It supports Food and Beverage by ensuring kitchen equipment, extraction, refrigeration, water, drainage and electricity function properly. It supports Security through fire panels, CCTV infrastructure, access systems and emergency systems. It supports Finance by controlling utilities, repairs, maintenance planning and CAPEX requests.

This cross-functional role explains why engineering should have direct access to senior leadership. If the Engineering Manager is isolated or treated only as a repair supervisor, critical risks may not reach the executive committee in time. A hotel's engineering condition affects operations, finance, guest satisfaction, safety, insurance and asset value. Therefore, engineering must be integrated into management discussions, not called only when something is already broken.

The Cost Center Misconception

A common misconception is that engineering is only a cost center because it spends money rather than generating revenue. This view is understandable but incomplete. Engineering payroll, spare parts, contracts, inspections and repairs appear as costs in the operating statement. However, the value of engineering is often measured by what does not happen: rooms do not go out of order, guests do not complain, systems do not fail, emergency repairs do not explode the budget, and the asset does not deteriorate prematurely.

A better way to understand engineering is through risk-adjusted value. A preventive maintenance contract may seem expensive until compared with the cost of a breakdown during peak occupancy. A spare pump may seem unnecessary until the main pump fails on a weekend. A specialist inspection may seem like a technical formality until an insurer asks for evidence after an incident. Engineering expenditures should therefore be evaluated not only by their direct cost, but also by the risk they reduce.

Engineering also contributes to margin through energy management. Heating, cooling, lighting and power consumption influence operating costs every day. Small improvements in settings, maintenance, monitoring and equipment performance can produce significant savings over time. These savings may be less visible than room revenue, but they strengthen profitability. In this sense, engineering can protect both revenue and margin.

When managers underinvest in engineering, they often create hidden debt. Deferred maintenance is not the same as savings. It is a transfer of cost into the future, usually with interest. A postponed repair can become a larger repair. A missed inspection can become a compliance issue. A minor guest complaint pattern can become a brand problem. A technical defect can become an emergency capital project. Strong managers recognize that prevention costs less than crisis.

The Five Core Missions of the Engineering Department

1. <u>Prevent Breakdowns</u>

The first mission of engineering is prevention. Preventive maintenance is the systematic care of equipment and systems before they fail. It includes inspections, cleaning, testing, lubrication, calibration, replacement of worn parts, monitoring and documentation. The objective is not to eliminate every failure, which is impossible, but to reduce avoidable failures and detect early signs of deterioration.

Preventive maintenance requires discipline because its results are often invisible. When it works, nothing happens. That is precisely its value. A preventive culture is different from a reactive culture. In a reactive culture, the department waits for complaints, alarms or breakdowns. In a preventive culture, the department uses schedules, data, checklists, technician observations and historical patterns to anticipate problems. Future managers must learn to ask whether engineering is mainly reacting or preventing.

2. Maintain Guest Comfort

The second mission is guest comfort. Comfort includes room temperature, air quality, lighting, water pressure, hot water availability, noise control, elevator reliability, internet infrastructure coordination and the general physical condition of the property. Many comfort issues are technical, but guests experience them emotionally. A room that is too hot can become frustration. A shower with weak pressure can become disappointment. A light that does not work can become a perception of neglect.

Comfort also depends on response time. Even well-maintained hotels experience defects. What matters is how quickly and professionally the issue is resolved. Engineering must work closely with Front Office and Housekeeping to prioritize guest-impacting issues. A small defect in a vacant room is different from a defect in an occupied VIP suite. Managerial judgment is needed to decide urgency, communication and compensation strategy.

3. Ensure Legal and Safety Compliance

The third mission is compliance. Engineering plays a major role in fire systems, emergency lighting, elevators, electrical safety, water systems, gas installations, pressure equipment and other regulated areas depending on the jurisdiction. Compliance requires more than technical action. It requires documentation, inspection planning, certificates, contractor records, corrective action follow-up and management visibility.

A manager should never assume that compliance exists simply because systems appear to work. A fire alarm panel may show no alarm, but testing records may be incomplete. An elevator may operate normally, but inspection documentation may be missing. A water system may deliver hot water, but temperature controls may create risk. Compliance is therefore both a technical and administrative discipline. The hotel must be able to prove that it manages safety responsibly.

4. Protect the Asset

The fourth mission is asset protection. Engineering teams preserve the long-term condition of the building and equipment. This includes guestrooms, public areas, technical rooms, roofs, facades, kitchens, laundries, spas, pools, back-of-house areas, external spaces and building systems. Asset protection requires lifecycle thinking: every system has a condition, an expected life, a replacement horizon and a maintenance requirement.

A hotel that looks beautiful at the front of house may still have serious hidden engineering risk. Plant rooms, risers, electrical rooms, drainage systems and roofs may reveal the true condition of the asset. Managers should understand that cosmetic renovation without technical investment can create a dangerous imbalance. A refreshed lobby does not compensate for weak infrastructure. Asset value depends on both visible design and invisible reliability.

5. Support Operations

The fifth mission is operational support. Engineering enables departments to deliver service without infrastructure obstacles. This includes responding to maintenance requests, preparing rooms, supporting events, coordinating with contractors, managing technical shutdowns, assisting with renovation works, supporting sustainability initiatives and communicating risks to management.

Operational support does not mean that engineering should be reduced to a helpdesk. The department must balance immediate requests with preventive work and long-term planning. If all technicians are constantly pulled into reactive tasks, preventive maintenance will suffer. If preventive maintenance is ignored, reactive tasks will increase. The manager's role is to protect the right balance between daily service support and strategic prevention.

Visible and Invisible Engineering

Visible engineering is what guests and employees directly experience: room temperature, lighting, hot water, elevators, functioning plugs, clear signage, safe public spaces and reliable equipment. Invisible engineering is what makes those outcomes possible: chillers, boilers, pumps, valves, pipes, electrical panels, control systems, ventilation units, fire panels, emergency generators, preventive maintenance schedules and contractor inspections.

This distinction is important because managers often receive feedback only about visible outcomes. A guest says the room is cold; the manager must investigate whether the cause is a thermostat, fan coil unit, valve, chiller setting, air balance issue, window insulation or user misunderstanding. A restaurant says the kitchen is too hot; the cause may involve extraction, make-up air, equipment heat load or maintenance. A conference organizer says the room is uncomfortable; the cause may be occupancy load, air-conditioning zoning or poor pre-event planning.

Managers do not need to diagnose every technical cause alone, but they must understand that visible symptoms can have invisible roots.

Good leadership means asking questions that connect symptoms to systems. How many rooms are affected? Is this isolated or repeated? Has this happened before? What is the temporary solution? What is the root cause? What is the cost of permanent correction? What is the guest impact? What is the risk if we wait? These questions move the discussion from complaint handling to engineering governance.

Engineering as Risk Management

Engineering risk can be grouped into four categories: operational, safety, financial and legal. Operational risk concerns the ability of the hotel to function. It includes room outages, equipment breakdowns, kitchen disruptions, elevator failures, water leaks, power issues and system interruptions. Safety risk concerns harm to guests, employees or visitors. It includes fire protection, emergency systems, electrical hazards, water quality, gas safety and physical hazards. Financial risk concerns repair costs, replacement costs, compensation, lost revenue, energy waste and unexpected capital expenditure. Legal risk concerns regulatory non-compliance, insurance exposure, contractual failures and liability.

These risks are interconnected. A legal risk can become financial. A financial decision can increase operational risk. An operational failure can become reputational. A safety issue can become a brand crisis. Engineering departments therefore manage more than equipment; they manage a network of risks affecting every stakeholder.

The General Manager's role is not to replace technical experts but to ensure that these risks are visible, prioritized and acted upon. A good GM asks for risk registers, maintenance plans, critical equipment lists, compliance calendars, incident reports, energy data and CAPEX forecasts. A weak GM waits for emergencies and then asks why nobody warned the executive committee. Engineering leadership requires governance, not improvisation.

Who Suffers When Engineering Fails?

When engineering fails, guests suffer first because they experience discomfort, inconvenience or safety concern. Their room may be too hot, the shower may not work, the elevator may be unavailable, the alarm may trigger unnecessarily, or a leak may disturb their stay. Guest dissatisfaction can lead to compensation, negative reviews, complaints to the brand or loss of future business.

Staff also suffer. Front Office teams absorb complaints. Housekeeping teams must move guests, inspect rooms, report defects and work around blocked areas. Food and Beverage teams may lose equipment or service capacity. Security may manage emergency responses. Finance may process refunds. Sales may face event complaints. The emotional pressure of engineering failures often lands on employees who did not create the technical issue but must manage the guest reaction.

Management suffers because technical failure becomes a leadership problem. The executive committee must coordinate decisions, communicate with ownership, manage costs, protect reputation and restore confidence. If failures are repeated, management credibility weakens. Owners suffer because asset value, revenue, insurance position and capital planning are affected. In serious cases, engineering failures can influence brand audits, insurance premiums, legal claims, employee morale and owner confidence in the operator.

The cascade effect demonstrates why engineering cannot be treated as a back-of-house detail. A technical issue may start in a plant room, but its consequences reach the lobby, the guestroom, the restaurant, the balance sheet and the owner's boardroom.

Engineering's Position in the Hotel Organization

In many hotels, the Chief Engineer, Director of Engineering or Engineering Manager reports directly to the General Manager or to the Director of Operations depending on property size and brand structure. Direct access to senior leadership is important because engineering decisions often involve safety, budget, guest impact and asset risk. If the engineering leader is too far from decision-making, serious issues may be filtered, delayed or misunderstood.

Engineering also requires strong cross-functional links. With Housekeeping, engineering coordinates room maintenance, room status, preventive inspections, defect reporting and out-of-order decisions. With Front Office, engineering supports guest complaint resolution, emergency response and communication about room moves or service recovery. With Security, engineering coordinates fire systems, emergency lighting, access control, CCTV infrastructure and crisis response procedures. With Finance, engineering plans operating budgets, service contracts, spare parts, energy costs and capital expenditure. With Food and Beverage, engineering supports kitchen equipment, refrigeration, ventilation, drainage and event setups. With Human Resources, engineering contributes to training, safety procedures and contractor access rules.

The organizational position of engineering should reflect its strategic importance. The department should not be judged only by the number of work orders closed. It should be assessed by reliability, risk reduction, compliance status, response time, preventive maintenance completion, guest-impacting defects, energy performance and quality of planning. A well-positioned Engineering Department gives leadership early visibility on risk.

Engineering Team Structure

Engineering team structure varies by hotel size, market level, building complexity, brand standards, local regulation and outsourcing strategy. A small limited-service hotel may have a lean maintenance team with external contractors for specialized systems. A large luxury hotel may have a Director of Engineering, assistant engineers, supervisors, shift technicians, painters, carpenters, plumbers, electricians, HVAC specialists, pool technicians, energy managers, administrative support and contracted specialists. Resort properties may also need landscape, water treatment, beach or pool infrastructure expertise.

A typical structure includes a Chief Engineer or Director of Engineering who leads the department, plans budgets, manages compliance, coordinates contractors, reports to senior management and supervises the team. An Assistant Engineer or Engineering Supervisor may coordinate daily operations, shifts, preventive maintenance and work orders. Technicians perform routine maintenance, repairs, inspections and guestroom interventions. Specialists may focus on HVAC, plumbing, electrical systems, kitchen equipment, refrigeration, fire systems, elevators or building management systems. External contractors provide certified expertise, complex repairs, statutory inspections or specialized maintenance.

The right structure is not identical for every hotel. It must be scaled to the risk profile of the property. A high-rise hotel with complex elevators, large meeting spaces and sophisticated mechanical systems needs different capabilities from a small boutique property.

A resort with pools, villas and extensive grounds has different needs from an urban business hotel. The managerial question is not "How many engineers should we have?" but "What capabilities are required to protect operations, guests and the asset at this property?"

Generalists Versus Specialists

Generalist technicians are multi-skilled employees who can handle a broad range of daily issues: replacing fixtures, addressing minor plumbing defects, checking lighting, responding to room maintenance calls, supporting basic equipment troubleshooting and performing routine preventive tasks. Their strength is flexibility. They are especially valuable in daily operations because they can respond quickly to varied requests and reduce dependency on external contractors for minor issues.

Specialists have deeper expertise in specific systems such as HVAC, electrical distribution, fire systems, elevators, water treatment, kitchen equipment, building automation or energy management. Their strength is precision and technical depth. Some systems require certified specialists because of safety, regulation or complexity. A generalist may identify that an elevator is out of service, but an elevator contractor may be required to intervene. A hotel technician may reset a system, but a fire alarm specialist may be needed for testing, certification or complex fault diagnosis.

The managerial decision is to balance fixed cost, response speed, risk and expertise. Employing many specialists internally can improve control and knowledge but increases payroll. Relying heavily on contractors reduces fixed payroll but increases dependency and may slow response. The best model is usually blended: internal generalists for daily operations and institutional knowledge, supported by external specialists for complex, regulated or high-risk systems. The right balance depends on property complexity, brand positioning, occupancy, location and risk tolerance.

Internal Team Versus Outsourcing

The decision between an internal team and outsourcing is one of the most important managerial questions in engineering. An internal team offers faster response, stronger knowledge of the property, better control over priorities and closer integration with operations. Internal employees understand recurring defects, guest expectations, brand standards, back-of-house realities and the informal knowledge of the building. They can respond quickly to occupied-room problems, event emergencies and urgent operational requests.

Outsourcing offers access to specialized expertise, variable cost, certified intervention and technical depth that may not be justified as full-time payroll. Contractors are essential for elevators, fire systems, major HVAC works, statutory inspections, complex electrical tasks, specialized kitchen equipment, water treatment, building automation and major renovations. Outsourcing can be efficient when the need is periodic, highly specialized or legally regulated.

However, outsourcing also transfers some operational dependency to external parties. Response times may be slower. Contractors may not understand guest impact. Knowledge can remain outside the hotel. Costs can increase during emergency calls. The hotel may become dependent on vendors for critical systems. Most importantly, outsourcing execution does not outsource accountability. If a contractor fails to maintain a system properly, guests and regulators will still look to the hotel's management. The hotel remains responsible for selecting, supervising and documenting contractor performance.

A useful executive question is: "What risk am I outsourcing?" If the risk is low and the expertise need is occasional, outsourcing may be logical. If the risk is high, response time critical, or guest impact immediate, the hotel may need stronger internal capability or a more robust service agreement. Outsourcing is not a way to ignore engineering. It is a governance choice that must be actively managed.

Engineering Hierarchy and Accountability

Hierarchy in engineering matters because technical decisions often involve safety, cost, urgency and operational impact. Without clear hierarchy, teams may hesitate, duplicate work, miss priorities or fail to escalate. In emergencies, unclear authority can be dangerous. Who can shut down a system? Who can call an emergency contractor? Who can approve overtime? Who communicates with guests? Who informs the General Manager? Who decides whether rooms are taken out of order? These questions must be answered before the crisis.

Accountability is different from execution. A technician may execute a repair. A contractor may perform an inspection. A supervisor may coordinate a shift. The Chief Engineer may lead the department. But accountability for the hotel's performance ultimately reaches senior management. The General Manager may delegate responsibility, but cannot fully delegate accountability. If engineering failure affects guest safety, legal compliance, asset condition or brand reputation, the GM will be expected to explain how the risk was managed.

Clear hierarchy prevents three common problems. First, it reduces safety incidents because authority for critical decisions is known. Second, it prevents delayed decisions because escalation paths are defined. Third, it controls budget overruns because spending authority and approval processes are established. Hierarchy should not create bureaucracy for its own sake. It should create clarity, speed and accountability.

The Role of the General Manager

The General Manager's role in engineering is to connect technical reality with business responsibility. The GM does not need to know every specification, but must know which engineering issues threaten guests, operations, compliance, reputation or asset value. The GM should ask questions that reveal root causes rather than only symptoms.

For example, when told that air-conditioning has failed in several rooms, the GM should ask: How many rooms are affected? Are they occupied or vacant? Is the issue isolated or systemic? What is the temporary solution? What is the expected time to resolution? What is the guest communication plan? Is there a repeated pattern? What preventive maintenance has been done? Is CAPEX needed? What is the risk if we delay? These questions do not require engineering expertise; they require managerial discipline.

The GM must also arbitrate budget trade-offs. Engineering often competes with visible projects such as decoration, marketing or guest-facing upgrades. Yet some invisible investments are essential. Replacing a failing pump, upgrading a fire panel, repairing a roof, improving ventilation or modernizing electrical distribution may be less attractive than redesigning a lobby, but it may protect the hotel more effectively. A strong GM understands the difference between cost avoidance and risk accumulation.

Finally, the GM sets the culture. If leadership only celebrates emergency heroics, the organization may remain reactive. If leadership values prevention, documentation, root-cause analysis, planned maintenance and cross-functional communication, engineering becomes strategic. The GM must make engineering visible before failure forces visibility.

Management Practices That Strengthen Engineering Governance

To manage engineering effectively, hotel leaders should establish a small number of practical routines. The first is a weekly engineering review that includes major defects, out-of-order rooms, preventive maintenance completion, safety issues, contractor works and guest-impacting problems. The second is a monthly risk and asset review that examines recurring failures, compliance status, critical equipment, energy consumption and CAPEX needs. The third is a clear escalation protocol for emergencies and high-impact failures.

Managers should also require accurate documentation. Work orders should be recorded and closed with meaningful information, not vague notes. Preventive maintenance should be scheduled and tracked. Contractor reports should be stored. Certificates should be accessible. Repeated defects should be analyzed. Energy and utility performance should be monitored. Documentation transforms engineering from informal problem-solving into professional asset management.

Another useful practice is cross-department training. Front Office and Housekeeping should know how to report defects accurately. Engineering should understand guest priority and service language. Security should understand engineering interfaces with emergency systems. Finance should understand the difference between maintenance, repair, CAPEX and deferred risk. When departments understand each other, technical issues are resolved faster and with less confusion.

Finally, engineering governance requires communication with ownership. Owners need clear explanations of risk, cost, asset condition and investment priorities. A technical problem should be translated into business language: guest impact, revenue risk, safety exposure, expected cost, timeline, options and recommendation. This translation skill is one of the most important capabilities future hospitality managers can develop.

In summary

The Engineering Department is one of the most important but least visible functions in a hotel. Its work supports business continuity, guest comfort, safety, compliance, asset protection and cost control. When engineering performs well, guests rarely notice it. When it fails, the consequences quickly spread across departments and become operational, financial, reputational and legal issues.

Engineering should not be reduced to repairs. Its most valuable contribution is prevention. Preventive maintenance, clear reporting, risk monitoring, documentation and lifecycle planning protect the hotel from avoidable crises. Strong engineering management keeps rooms sellable, equipment reliable, safety systems compliant and the owner's asset protected.

The chapter also showed that engineering supports every department. Front Office, Housekeeping, Food and Beverage, Security and Finance all depend on reliable infrastructure. This is why engineering must be connected to senior leadership and integrated into executive decision-making. It is not only a technical department; it is a strategic support function.

Engineering teams can be structured in different ways depending on property size, complexity and risk. Managers must understand the trade-offs between generalists and specialists, internal teams and outsourced contractors. A blended model is often appropriate, but outsourcing execution does not remove hotel accountability.

Finally, the role of the General Manager is essential. The GM must ask the right questions, understand risk, support preventive investment, clarify accountability and ensure that engineering decisions align with guest safety, brand standards, financial discipline and asset value. In hospitality, engineering failure is never only an engineering problem. It is a leadership problem.

OPERATIONAL CASE STUDY

The Silent Breakdown at Hotel Aurélia Grand

Hotel Aurélia Grand is a 210-room upscale hotel located in a major European capital. The property includes two restaurants, a rooftop bar, six meeting rooms, a small spa, three guest elevators, one service elevator, and a busy conference business during weekdays. The building is twenty-five years old and has undergone several cosmetic renovations, but many technical systems are original or only partially upgraded. The hotel is positioned as elegant, reliable and business-friendly. Its average occupancy is 82%, with frequent peaks above 95%.

For several months, the hotel has experienced recurring technical issues. Guests occasionally complain that rooms are too warm at night. Housekeeping reports that several rooms have repeated lighting faults. The rooftop bar team complains about inconsistent ventilation during busy evenings. The Front Office Manager notes that room moves due to technical defects have increased, but the issues are usually solved one by one and not discussed at executive committee level. The Chief Engineer has warned informally that preventive maintenance is behind schedule because the team is short-staffed and constantly responding to urgent requests.

The Engineering Department consists of one Chief Engineer, one assistant engineer, four generalist technicians and several external contractors. There is no dedicated HVAC specialist on property. Elevator maintenance, fire systems, kitchen equipment and major HVAC works are outsourced. The Chief Engineer reports to the Director of Operations, who reports to the General Manager. The GM receives monthly financial results but no structured engineering risk report. Preventive maintenance completion is not regularly reviewed by the executive committee.

During a fully booked week, the hotel hosts a three-day pharmaceutical conference with 280 delegates. On the first evening, one guest elevator stops working during the dinner rush. The elevator contractor is called but cannot attend until the next morning. Guests complain about long waiting times. Housekeeping struggles to deliver amenities and prepare rooms. The following morning, several meeting rooms become too warm during a plenary session. The conference organizer complains that the environment is unacceptable for a premium event. Engineering resets part of the system, but the temperature problem returns after lunch.

At 2:15 a.m. on the second night, air-conditioning fails in thirty-six rooms on the upper floors. The night manager receives repeated calls from angry guests. Some guests request refunds. Others ask to move rooms, but the hotel is nearly full. The night engineer discovers that a key HVAC component has failed and that a temporary repair may not hold. The Chief Engineer, contacted by phone, explains that this component had been mentioned in previous maintenance notes but replacement was postponed because the department was trying to stay within budget.

The next morning, the General Manager faces several problems at once: angry guests, a dissatisfied conference organizer, one elevator still out of service, pressure from the owner's representative to control costs, and a Chief Engineer requesting urgent approval for external specialists and replacement parts. Online reviews have already appeared mentioning "poor maintenance" and "not worth the price." The Front Office team feels abandoned. The Engineering team feels blamed for problems they had warned about. Finance asks whether the costs should be treated as operating expense or emergency CAPEX.

Case Questions for Discussion

1. What were the early warning signs that the hotel's engineering risk was increasing?
2. Which failures were technical, and which were managerial or organizational?
3. What questions should the General Manager ask within the first hour of the crisis?
4. How should the hotel stabilize guest experience during the next 48 hours?
5. What information should be communicated to the owner's representative?
6. Should the hotel increase internal engineering capability, strengthen outsourcing contracts, or both?
7. How should preventive maintenance be made visible at executive committee level?
8. What would you change in the reporting line, meeting structure or KPI dashboard?
9. How should the hotel evaluate the cost of prevention compared with the cost of crisis?
10. What leadership lesson should the General Manager take from this event?

Final learning note: Expected Managerial Analysis

The core issue is not only the failed HVAC component or the broken elevator. The deeper issue is weak engineering governance. Preventive maintenance was behind schedule, technical warnings were informal, executive visibility was limited, and budget discipline appears to have been interpreted as postponement rather than risk management. The hotel treated engineering as a support cost until engineering became a business crisis.

Students should identify that the General Manager needs to act on two levels: immediate stabilization and structural correction. Immediate actions include guest recovery, room allocation decisions, emergency contractor support, temporary equipment if available, clear communication with the conference organizer, and support for front-line employees. Structural actions include an engineering audit, a preventive maintenance recovery plan, clearer reporting to executive committee, review of outsourcing service-level agreements, CAPEX prioritization and stronger risk communication to ownership.

The strongest answer recognizes that accountability goes upward. The Chief Engineer may manage technical execution, but senior leadership must create the conditions for prevention. If the GM does not receive engineering risk information, the reporting system must change. If the information exists but is ignored, leadership discipline must change. If budget pressure leads to repeated deferral, risk-adjusted financial decision-making must change.

CHAPTER KEY TERMS

- **Engineering Department:** The hotel function responsible for maintaining infrastructure, systems, equipment, safety support, technical reliability and business continuity.
- **Business Continuity:** The ability of the hotel to continue operating without major interruption despite technical risks or incidents.

- **Preventive Maintenance:** Planned maintenance performed before equipment fails in order to reduce breakdowns, extend asset life and protect operations.
- **Reactive Maintenance:** Maintenance performed after a defect, complaint or breakdown has already occurred.
- **Asset Protection:** The systematic care of the hotel building and equipment to preserve value, safety and long-term performance.
- **Chief Engineer / Director of Engineering:** The manager responsible for leading the engineering function, coordinating technical teams, managing risk, supporting compliance and advising senior leadership.
- **Generalist Technician:** A multi-skilled technician able to handle a broad range of routine maintenance tasks and minor repairs.
- **Specialist Technician:** A technician or contractor with deep expertise in a specific system such as HVAC, electrical, fire systems or elevators.
- **Outsourcing:** The use of external contractors or service providers to perform specific technical tasks, maintenance contracts or specialist interventions.
- **Accountability:** The obligation to answer for decisions, outcomes and risk management, even when execution is delegated.
- **Compliance:** The condition of meeting legal, safety, insurance, brand or regulatory obligations supported by proper documentation.
- **Risk-Adjusted Value:** The value of an engineering decision when both direct cost and risk reduction are considered.
- **Lifecycle Thinking:** A long-term view of equipment and building systems, including maintenance, replacement, upgrades and total cost over time.

- **CAPEX:** Capital expenditure used for investments that improve, replace or extend the life of an asset.
- **OPEX:** Operating expenditure used for day-to-day running costs, routine maintenance, payroll, utilities and service contracts.
- **Work Order:** A documented maintenance request or task that records the issue, priority, action taken and closure status.
- **Critical Equipment:** Equipment whose failure would have major impact on safety, guest comfort, revenue, compliance or operations.
- **Service-Level Agreement:** A formal agreement defining expected contractor performance, response times, scope, responsibilities and standards.
- **Escalation Protocol:** A defined process indicating when and how issues should be elevated to higher levels of management.
- **Engineering Governance:** The leadership routines, reporting structures, documentation and decision-making processes that ensure engineering risks are managed properly.

CHAPTER REVIEW QUESTIONS

- Why should the Engineering Department be considered a business continuity function rather than only a maintenance function?
- Explain how an engineering failure can move from a technical issue to a guest experience, financial and reputational problem.

- What are the five core missions of a hotel Engineering Department, and why does each matter to a General Manager?
- Why is preventive maintenance often financially stronger than reactive maintenance, even when it appears to cost more in the short term?
- How does engineering support Front Office, Housekeeping, Food and Beverage, Security and Finance?
- What are the advantages and disadvantages of relying mainly on internal engineering employees?
- What are the advantages and risks of outsourcing engineering work to external contractors?
- Why is the distinction between execution and accountability important in hotel engineering?
- What engineering information should a General Manager expect to see regularly at executive committee level?
- In the case study, what were the main governance failures, and what corrective actions should the hotel implement?

CHAPTER REFERENCES

- ASHRAE. ASHRAE Handbook series: HVAC Applications; Fundamentals; HVAC Systems and Equipment. American Society of Heating, Refrigerating and Air-Conditioning Engineers.
- Borsenik, F. D., & Stutts, A. T. The Management of Maintenance and Engineering Systems in the Hospitality Industry. John Wiley & Sons.

- International Facility Management Association (IFMA). Facility management body of knowledge and professional practice resources.
- International Organization for Standardization. ISO 41001: Facility management - Management systems - Requirements with guidance for use.
- International Organization for Standardization. ISO 55000: Asset management - Overview, principles and terminology.
- National Fire Protection Association. NFPA 101: Life Safety Code and related fire protection standards.
- Ransley, J., & Ingram, H. Developing Hospitality Properties and Facilities. Routledge.
- RICS. Hotel and real estate asset management guidance and professional resources.
- Strategic hotel operations and engineering management materials from the course presentation “Engineering Department Fundamentals” and the accompanying professor notes.

Chapter 2

Hotel Building Infrastructure and Technical Systems

Understanding What Makes a Hotel Function

CHAPTER GENERAL OVERVIEW

A hotel can be understood as a living organism. It breathes through its ventilation systems, circulates energy through electrical distribution, moves water through pumps and pipes, regulates temperature through heating and cooling, protects life through emergency systems, and manages waste, air quality and sustainability through daily technical routines. Guests rarely see this organism directly. They experience its effects: a room that feels fresh, a shower that has pressure and hot water, a corridor that is well lit, an elevator that arrives smoothly, a breakfast area that remains comfortable during peak time, and a building that feels safe at all hours.

This chapter explores the physical reality behind guest comfort. It is not designed to make hospitality students become mechanical, electrical or plumbing engineers. Its purpose is managerial. Future hotel managers must understand how technical systems shape guest experience, risk exposure, financial performance and asset value. They must know which questions to ask, how to connect technical failures to business consequences, and how to participate in investment decisions without pretending to be technical specialists.

The key idea is simple: infrastructure equals guest experience. The guest may describe a problem emotionally - “I did not sleep well”, “the room felt stuffy”, “the shower was weak”, “the air conditioning was noisy”, “the hotel did not feel fresh” - but behind each complaint there is usually a technical system. HVAC, plumbing, electricity, lighting, emergency power, ventilation, building controls, waste management and air quality are not background topics. They are part of the service promise.

Infrastructure also represents risk. A technical system that performs poorly can affect guest satisfaction, online reputation, regulatory compliance, insurance exposure, energy costs and long-term asset value. A hotel is not a static product. It is a building that operates continuously, with guests, staff, contractors and equipment using the same infrastructure every day and night. A failure in one system can quickly become an operational crisis. A badly maintained ventilation system can become a guest-comfort issue. A weak hot-water system can become a refund issue. A faulty generator can become a life-safety issue. Poor zoning and controls can become an energy-cost issue.

The chapter therefore develops an executive view of infrastructure. It introduces the main technical systems that make hotels function: HVAC, plumbing and water management, electrical systems and emergency power, heat, light and power distribution, IoT and smart controls, waste management, sustainability constraints, and CO2 monitoring for indoor air quality. For each topic, the emphasis is not on detailed engineering design. The emphasis is on operational dependency, guest impact, cost, risk and decision-making.

A future General Manager, Operations Director, Asset Manager or Rooms Division leader does not need to repair a boiler or design an air-handling unit. However, that manager must be able to ask: What happens if this system fails? How many rooms or outlets are affected? Is the failure isolated or systemic? Is preventive maintenance funded? What is the energy impact? What is the guest impact? What is the compliance exposure? What investment options exist, and what are the trade-offs between CAPEX, OPEX, guest satisfaction and risk?

Managerial principle: You do not need to fix the system. You need to understand the consequences, ask the right questions, and decide on risk.

CHAPTER CONTENT

A Hotel as a Living Organism

The idea that a hotel is a living organism helps managers understand interdependence. No department operates in isolation. Front Office depends on elevators, door locks, PMS access, power, lighting, air-conditioning and guestroom readiness. Housekeeping depends on laundry systems, water, vacuum equipment, ventilation and room maintenance. Food and Beverage depends on refrigeration, extraction, dishwashing, cooking equipment, water temperature, waste storage and safe electrical supply. Spa and wellness facilities depend on water quality, heating, humidity control, ventilation and lighting. Meetings and events depend on power, air-conditioning, lighting, audiovisual systems and safety systems.

When the organism is healthy, the hotel feels effortless. Guests do not think about the pump that creates water pressure, the plant room that produces chilled water, the distribution board that feeds the corridor lighting, or the building management system that regulates energy consumption. They simply experience comfort. The more reliable the infrastructure, the more natural the service feels. In this sense, good engineering allows hospitality to appear human, warm and spontaneous.

When one vital system stops, the organism reacts. An air-conditioning failure affects sleep quality, guest patience, staff workload, compensation requests and online reviews. A water-pressure problem affects showers, laundry, kitchens and housekeeping. A power outage affects elevators, payment systems, electronic locks, lighting, fire systems and operational communication. A drainage issue affects hygiene, odor, room availability and reputation. Technical failures therefore travel beyond the plant room. They move into guest emotions, staff stress, management decisions and financial results.

For students, the first shift is to move from facility thinking to business risk thinking. Facility thinking asks, “What equipment is broken?” Business risk thinking asks, “What is the operational exposure if this system underperforms, and how should management respond?” Both questions matter, but the second belongs to the hospitality manager. The engineer diagnoses the technical cause. The manager must understand the business effect and decide what level of risk is acceptable.

Infrastructure as Guest Experience

Hospitality often presents itself through visible details: design, service gestures, amenities, food presentation, uniforms, lighting mood and staff attentiveness. Yet many of the strongest guest impressions depend on infrastructure. A guest who says “I slept well” may be responding to the mattress, but also to room temperature, acoustic insulation, air circulation and humidity. A guest who says “the shower was excellent” is responding to plumbing, hot-water production, pressure regulation and drainage. A guest who says “the room felt fresh” is responding to ventilation, filtration and CO2 control. A guest who says “I felt safe and calm” is responding to emergency lighting, reliable elevators, stable power, controlled access and building integrity.

The guest usually cannot name the system behind the feeling. This creates a perception gap. The guest evaluates the outcome, not the infrastructure. Management must understand the system that creates the outcome. If the room is too hot at night, the guest may blame the hotel generally, not the chiller, fan coil, thermostat, insulation or maintenance schedule. If the corridor smells unpleasant, the guest may blame cleanliness, even when the root cause is drainage, ventilation or waste storage. If the elevator fails, the guest does not distinguish between an external contractor, a maintenance contract and an internal engineering team. The hotel owns the experience.

This is why infrastructure does not usually create delight in the same visible way as a welcome amenity or personalized service. Rather, infrastructure creates the conditions in which delight becomes possible. Its failure creates disaster. One technical discomfort can cancel several positive service moments because it touches the guest's body directly: heat, cold, noise, smell, water, darkness, air quality and safety. These are physical sensations before they become opinions. Once they become opinions, they quickly influence online reviews and future bookings.

Managerial Infrastructure Map

Guest perception	**Technical system behind it**	**Managerial question**
I slept well	HVAC, insulation, acoustics, lighting controls	Are temperature, noise and air quality monitored by room or by zone?

The shower was pleasant	Plumbing, hot-water production, pressure regulation, drainage	Can the system meet peak demand in mornings and evenings?
The room felt fresh	Ventilation, filtration, CO2 monitoring, humidity control	Is air quality measured or only assumed?
The hotel felt safe	Emergency power, fire systems, lighting, elevators, access control	Which systems remain operational during main power failure?
The stay felt seamless	Preventive maintenance, work-order response, smart controls	Are recurring defects analyzed before they become reputation problems?

Infrastructure as Risk Exposure

Infrastructure is not only a comfort issue. It is also a risk issue. A hotel's technical systems create five major categories of exposure: guest satisfaction, safety and compliance, energy cost, asset value and brand reputation. These categories are connected. A technical failure may begin as a comfort complaint but develop into a reputational or financial problem. A maintenance delay may begin as a budget saving but develop into asset deterioration. A compliance weakness may remain hidden until an inspection, incident or insurance review reveals it.

Guest satisfaction risk is the most visible. Temperature complaints, poor water pressure, bad smells, noise and power interruptions are among the fastest ways to damage guest perception. Unlike some service errors, technical failures are often difficult to compensate through friendliness alone. A receptionist can apologize for a room that is too hot, but cannot change the physical discomfort unless Engineering solves the underlying issue. Repeated technical complaints also erode staff confidence, because guest-facing employees must absorb frustration they cannot personally fix.

Safety and compliance risk is less visible but more severe. Emergency lighting, fire detection, alarms, smoke control, generators, evacuation systems, electrical panels, water treatment and elevator maintenance are not optional. They are part of the hotel's duty of care. When life-safety systems fail, the consequences can include evacuation, legal liability, insurance complications, regulatory sanctions and deep reputational damage. The managerial question is not whether the system is technically interesting; it is whether the hotel can prove that critical systems are maintained, tested and documented.

Energy cost risk is increasingly important. Hotels consume energy continuously. Heating, cooling, ventilation, lighting, laundry, kitchens, hot water, elevators and pumps all contribute to utility expenses. Inefficient infrastructure drains margin every day, not only when something breaks. A badly zoned HVAC system, outdated lighting, poorly calibrated controls or insufficient insulation may silently reduce profitability for years. Managers must therefore understand infrastructure as an operating-cost driver, not only as a maintenance topic.

Asset value risk concerns the long-term condition of the property. Deferred maintenance can be tempting when budgets are tight because the consequences may not be immediate. However, building systems age. Pumps, boilers, chillers, distribution networks, electrical panels, controls, pipework and plant rooms all have lifecycle realities. When maintenance is postponed, the hotel may appear to save money temporarily while accumulating future CAPEX. Asset managers and owners look closely at these hidden liabilities because they affect valuation, investment planning and renovation strategy.

Brand reputation risk connects all these elements. A technical failure can become public quickly. Guests may post about room heat, broken elevators, water problems or poor air quality. Corporate clients may reconsider conference contracts if infrastructure is unreliable. Luxury guests may interpret technical inconsistency as lack of care. In a competitive hotel market, infrastructure failures become brand signals. They tell guests whether the hotel is well managed behind the scenes.

A hotel is not a static product. It is a machine running continuously, and every system carries operational, financial and reputational consequences.

Chief Engineer, Director of Engineering and the Managerial Boundary
A useful distinction in hotel engineering is the difference between technical expertise and executive decision-making. The Chief Engineer or technical leader understands systems in depth. This role diagnoses failures, organizes maintenance, supervises technicians, coordinates contractors and ensures that equipment performs as expected. In some properties, the title Director of Engineering is used for a more senior role that combines technical leadership with strategic planning, budgeting, CAPEX prioritization, sustainability performance and ownership communication.

For hospitality managers outside the engineering function, the important lesson is to respect the technical boundary while owning the business decision. A Rooms Division Manager does not need to calculate cooling loads. A General Manager does not need to design a pump system. An Asset Manager does not need to repair an electrical panel. However, all three must understand the risks and choices presented by engineering information. They must decide whether to fund preventive maintenance, whether to approve equipment replacement, whether to close rooms, whether to compensate guests, whether to accelerate CAPEX and whether to accept or mitigate risk.

The phrase "You do not need to fix it. You need to decide" captures this boundary. Deciding requires enough knowledge to challenge vague explanations. A manager should not accept "the system is old" as a complete answer. The next questions should be: What is the failure pattern? What is the immediate impact? What is the root cause? What are the options? What are the costs? What happens if we delay? What guest areas are exposed? What compliance issues exist? What does the preventive maintenance record show? How does this affect revenue, safety and reputation?

Good engineering leadership translates technical reality into managerial language. Good general management translates managerial priorities into operational support for Engineering. When both sides communicate well, the hotel can make balanced decisions. When communication fails, the organization becomes reactive: technicians fix symptoms, managers argue about cost, guests experience discomfort, and owners discover problems too late.

HVAC as the Backbone of Comfort

HVAC stands for heating, ventilation and air conditioning. In hotels, HVAC is one of the most important infrastructure systems because it directly affects comfort, sleep quality, air circulation, humidity, perceived freshness and energy consumption. It is also one of the most frequent sources of guest complaints. Guests may not use the term HVAC, but they understand its failures immediately: too hot, too cold, too dry, too humid, too noisy, stuffy, unpleasant smell or uncomfortable airflow.

HVAC is not only temperature control. Heating and cooling are only one part of the system. Ventilation brings fresh air into the building and removes stale air. Air circulation distributes conditioned air through rooms and public spaces. Humidity control affects comfort and perceived quality. Filtration improves indoor air quality. Controls determine how guests, staff and automated systems adjust conditions. If any part of this chain is poorly designed, maintained or calibrated, the guest experience suffers.

The managerial challenge is that HVAC failures can be both immediate and systemic. A single room may have a faulty thermostat or fan coil. A floor may have a distribution issue. A wing may suffer because of poor zoning. The entire property may be affected by a chiller failure, boiler issue, control-system problem or inadequate preventive maintenance. Managers should therefore avoid treating every HVAC complaint as isolated. Repeated comfort complaints may indicate deeper design or maintenance problems.

HVAC also has a significant cost dimension. Heating, cooling and ventilation are major energy consumers in hotels. A system that keeps guests comfortable but wastes energy can damage margins. A system that saves energy but creates discomfort can damage revenue and reputation. The managerial goal is not simply to reduce energy use; it is to maintain comfort efficiently. This is where smart controls, good zoning, preventive maintenance, insulation and accurate monitoring become important.

Because comfort complaints are highly emotional, HVAC must be managed proactively. Guests who cannot sleep because of heat, noise or poor air quality are unlikely to evaluate the hotel rationally.

They feel physically affected. They may request room moves, refunds or compensation. They may leave negative reviews. Staff then spend time managing dissatisfaction rather than delivering hospitality. Preventive HVAC management is therefore a guest-experience strategy as much as a technical requirement.

Centralized and Decentralized HVAC

Hotels may use different HVAC configurations. A centralized system conditions air or water from a central plant and distributes it to areas of the hotel. A decentralized or room-based system gives more localized control, often through individual units or smaller systems. Neither solution is universally superior. Each creates trade-offs involving energy efficiency, maintenance complexity, guest control, capital cost and failure risk.

Centralized systems can be efficient and appropriate for larger properties. They allow coordinated energy management, centralized maintenance and integration with building controls. However, they can create cascading risk. If a central component fails, many rooms or public areas may be affected at once. Central systems can also be complex to maintain, requiring specialized knowledge, strong preventive routines and clear contingency planning. Managers should understand the exposure: if this system stops, how many rooms are impacted?

Decentralized systems can give guests greater control and reduce the scale of failure. If one unit fails, the problem may affect one room rather than an entire wing. This can be operationally useful. However, room-based systems may consume more energy, create more noise, require more individual maintenance and produce inconsistent comfort across the property. They may also complicate sustainability objectives if controls are weak or guests leave systems running inefficiently.

The managerial question is not which system is perfect. There is no perfect solution. The question is which trade-off fits the property, guest expectations, climate, energy cost, building design, brand positioning and maintenance capacity. A luxury resort in a hot climate, a compact urban hotel and a heritage building may require different approaches. Managers must be able to discuss these choices with engineers, owners and designers using the language of risk, cost, comfort and asset strategy.

Managerial Questions for HVAC

Managers should learn to ask practical HVAC questions that connect technical performance to business impact. The first question is: What happens if it fails? This means identifying failure scenarios. Does the hotel lose cooling in one room, one floor, an entire wing, public areas or the whole building? Does failure affect guest comfort only, or also life safety, food storage, spa operations or event spaces?

The second question is: How many rooms or revenue areas are impacted? A problem in two rooms is operationally different from a problem in forty rooms during high occupancy. Managers must connect HVAC risk to rooms revenue, meeting revenue, food and beverage revenue, guest compensation and staff workload.

The third question is: Is preventive maintenance funded and completed? HVAC systems need filters, cleaning, calibration, inspections, seasonal preparation, contractor servicing and replacement planning. If preventive maintenance is repeatedly postponed, the hotel is not saving money; it is increasing the probability of complaints, energy waste and failure.

The fourth question is: What is the energy cost impact? An HVAC decision may increase or reduce OPEX significantly. A manager should ask whether the system is overcooling or overheating, whether zoning is efficient, whether occupied and unoccupied rooms are managed differently, whether controls are used properly and whether staff understand the system.

The fifth question is: Is guest control limited, balanced or excessive? Some systems restrict guest control to protect energy performance. Others give guests wide control but increase consumption. The right balance depends on brand promise and market positioning. A luxury guest may expect precise control, while the hotel still needs intelligent limits to avoid waste.

Plumbing and Water Management

Water is one of the most immediate dimensions of hotel comfort. Guests expect hot water quickly, stable pressure, clean fixtures, effective drainage and no unpleasant odors. They expect this during peak usage periods, especially mornings and evenings. A hotel can recover from many minor service imperfections, but a poor shower is remembered strongly because it affects the body directly and is part of the daily ritual of travel.

Plumbing and water management include hot and cold water distribution, pumps, pressure regulation, hot-water production and storage, drainage, wastewater removal and water treatment. In larger or more complex properties, systems must serve guestrooms, kitchens, laundries, spas, pools, staff areas, irrigation and fire systems. Demand is not constant. It peaks when many guests shower at similar times, when kitchens are active, when laundry operates, or when events increase restroom usage.

Hot water carries a strong energy dimension. Producing and storing hot water can be expensive. The hotel must provide enough capacity for guest satisfaction without wasting excessive energy. Storage systems may support peak demand but require careful temperature management and maintenance. Instant systems may reduce storage but have capacity and technical constraints. Managers do not need to design these systems, but they must understand the business trade-off between comfort, energy, risk and investment.

Water pressure is another critical issue. Poor pressure leads immediately to complaints. Excessive pressure can damage systems and fixtures. Pressure may vary by floor, time of day or demand pattern. The engineering team must monitor pumps, valves, regulators and distribution networks. Managers should recognize that a water complaint may not be a housekeeping issue. It may be a pressure, capacity, heating or distribution issue.

Drainage and odor control also matter. Bad smells in corridors, guestrooms, bathrooms, kitchens or waste areas can damage the perception of cleanliness and quality. Guests often interpret odor as poor hygiene, even when the root cause is technical. Proper drainage, trap maintenance, ventilation, grease management and waste storage are essential. Managers should ensure that departments report odors clearly and that Engineering investigates root causes, not only symptoms.

Water management also has a compliance and health dimension. Legionella risk management is a serious example. Hotels must manage water temperature, circulation, stagnation, monitoring and documentation according to applicable regulations and professional guidance. Failure in this area can create legal exposure and guest safety risk. Water is therefore not just comfort and cost. It is also health governance.

Electrical Systems and Emergency Power

Electrical systems support nearly every function in the hotel. Lighting, sockets, elevators, kitchen equipment, ventilation, IT systems, door locks, payment systems, security systems, fire systems, pumps and building controls all depend on reliable electrical supply. A hotel without power loses much more than light. It loses operational coordination, guest safety perception, service capacity and sometimes basic access control.

Electrical infrastructure includes incoming supply, distribution boards, circuits, panels, wiring, protective devices, lighting systems, emergency lighting, generators, uninterruptible power supply for critical systems and monitoring. In hotels, electrical reliability is especially important because the building is occupied continuously and guests may be asleep, showering, using elevators or relying on electronic locks at the moment a failure occurs.

Emergency power is non-negotiable. When main power fails, life-safety systems must remain operational. Emergency lighting, fire detection, alarm systems, smoke control where applicable, selected elevator operation, critical IT systems, PMS access, door-lock systems, communication systems and security functions may all require backup or continuity planning. Managers must know which systems are supported, for how long, under which conditions and how often they are tested.

A backup generator should not be viewed only as a cost. It is reputation insurance and life-safety support. A hotel that manages a power failure calmly, safely and professionally protects guest trust. A hotel that leaves guests in darkness, elevators blocked, locks malfunctioning or staff unable to communicate faces serious reputational and liability consequences. The difference between these scenarios is not only equipment; it is planning, testing, training and communication.

Electrical systems also involve capacity planning. Renovations, new kitchen equipment, spa additions, EV charging, event technology, laundry upgrades or additional IT infrastructure may increase electrical demand. Managers should ensure that engineering and ownership assess capacity before commercial promises are made. Selling a new service without understanding infrastructure capacity can create operational failure later.

Heat, Light and Power Distribution

In hotel operations, distribution often matters as much as production. It is not enough to produce heat, cooling, water or electrical power. These resources must reach the right spaces, at the right time, in the right quantity, with the right controls. Poor distribution creates waste, discomfort and complaints even when central equipment is functioning.

Zoning is one of the most important distribution concepts. A hotel contains spaces with different usage patterns: occupied rooms, vacant rooms, corridors, restaurants, kitchens, meeting rooms, offices, spa areas, back-of-house spaces and outdoor areas. Treating all zones the same wastes energy and reduces comfort. Meeting rooms may need high ventilation during events and lower conditioning when empty. Guestrooms may need different settings when occupied, vacant or being cleaned. Kitchens need extraction and cooling strategies that differ from lounges or corridors.

Bad zoning creates hidden cost. If heating, cooling or lighting is applied to spaces that do not need it, the hotel pays continuously for waste. If zones are poorly controlled, staff may create informal workarounds, such as opening doors, using portable heaters, overriding thermostats or leaving lights on. These actions may solve immediate discomfort but increase risk and cost. Smart distribution systems reduce these conflicts by matching energy use to actual demand.

Lighting is both operational and emotional. It affects safety, atmosphere, brand perception, wayfinding and energy cost. Poor lighting can make a corridor feel unsafe, a lobby feel cold, a restaurant feel unattractive or a back-of-house area unsafe for staff. LED systems, occupancy sensors, timers and lighting scenes can reduce energy use while improving experience when properly designed and maintained. However, lighting controls must remain understandable for staff and aligned with service needs.

Power distribution also affects operational resilience. Critical loads should be identified. The hotel must know which areas and systems are essential during disruption. Not every socket or decorative light needs backup power, but life safety, communication, access and selected operational systems require continuity planning. Managers should ask for clear maps of critical systems and ensure that emergency procedures match the actual technical design.

IoT and Smart Building Control

IoT, or the Internet of Things, refers to connected devices that measure, communicate and sometimes act automatically. In hospitality infrastructure, IoT can include occupancy sensors, smart thermostats, CO2 sensors, leak detectors, energy meters, equipment sensors, lighting controls, predictive maintenance alerts and building management systems. The managerial definition can be summarized as sensors plus data plus rules equals operational control.

The value of IoT is not the gadget itself. A sensor is only useful if the hotel knows what to do with the information. A smart thermostat is only valuable if it improves comfort or energy performance. A leak detector is only valuable if alerts reach the right person quickly. A predictive maintenance alert is only valuable if it triggers action before failure. Without operational ownership, IoT becomes an expensive layer of technology that produces data but not performance.

Smart building technology can support HVAC optimization by adjusting temperature and ventilation based on occupancy. It can reduce energy consumption in vacant rooms. It can monitor CO2 and improve air quality. It can detect abnormal equipment performance before breakdown. It can help identify energy waste. It can provide management with better visibility over patterns that were previously hidden.

However, smart technology also introduces managerial risks. Systems may be poorly integrated. Staff may not be trained. Data may be ignored. Vendors may oversell benefits. Privacy and cybersecurity questions may arise. Maintenance teams may become dependent on platforms they do not fully understand. Managers should therefore evaluate IoT investments through business questions: What operational problem does it solve? Who owns the data? Who responds to alerts? What savings or guest benefits are expected? What happens if the system fails?

The best smart building strategy is not technology enthusiasm. It is disciplined operational control. IoT should serve guest consistency, risk reduction, sustainability performance and margin protection. It should make the invisible building more visible to management without creating complexity that the organization cannot manage.

Waste Management and Sustainability

Waste management is often underestimated as an engineering and operational topic. Hotels generate waste from guestrooms, restaurants, kitchens, events, spas, offices, maintenance activities, packaging, landscaping and refurbishment works. Waste must be sorted, stored, moved, collected and documented while avoiding odor, pests, hygiene problems, guest exposure and regulatory issues.

A waste system is both technical and behavioral. It requires space, containers, ventilation, cleaning routines, collection schedules, staff training, supplier coordination and compliance awareness. If waste storage is poorly designed, odors may spread. If sorting is unclear, recycling targets fail. If logistics are inefficient, back-of-house corridors become cluttered. If staff are not trained, sustainability commitments remain theoretical. Managers should view waste as a daily operational system, not a secondary housekeeping issue.

Sustainability has also become an operational imperative. Hotels face increasing regulatory pressure, brand commitments, guest expectations, ESG reporting demands and rising energy costs. Sustainability is therefore not only a marketing message. It influences procurement, energy management, water consumption, waste reduction, maintenance, renovation planning, reporting and owner relations.

From an engineering perspective, sustainability includes efficient HVAC, LED lighting, smart controls, insulation, water-saving fixtures, heat recovery, renewable energy where feasible, preventive maintenance, waste reduction, monitoring and staff behavior. The challenge is to balance sustainability with guest satisfaction. A water-saving showerhead that damages perceived quality may create complaints. A temperature policy that saves energy but makes guests uncomfortable may hurt reputation. Good sustainability management finds solutions that reduce impact without weakening the hospitality promise.

The executive question is no longer whether sustainability matters. It does. The question is how to embed it in daily operations and investment decisions. Managers must distinguish symbolic actions from operational impact. They must also understand that some sustainability investments may require CAPEX but reduce OPEX, regulatory exposure and brand risk over time.

CO2 Monitoring and Indoor Air Quality

Indoor air quality is increasingly important in hospitality. Guests may not measure air quality, but they feel its effects. A room that feels stuffy, heavy or stale can reduce comfort and sleep quality. Poor ventilation may contribute to headaches, fatigue and dissatisfaction. Elevated CO2 levels can indicate insufficient fresh air relative to occupancy. In meeting rooms, restaurants, gyms and guestrooms, air quality is part of the experience.

CO2 monitoring provides a practical signal. It does not measure every aspect of air quality, but it helps indicate whether ventilation is adequate for the number of people in a space. Smart systems can adjust ventilation based on real-time occupancy and CO2 levels. This creates a balance between comfort and energy efficiency: ventilate when needed, reduce waste when spaces are empty or lightly occupied.

For managers, CO2 monitoring matters because it connects guest comfort, wellness, sustainability and operational control. A hotel may claim to offer wellness or premium sleep quality, but poor ventilation undermines that promise. A conference hotel may sell productive meeting environments, but a stuffy meeting room damages participant experience. A sustainability program may reduce ventilation to save energy, but if not controlled intelligently it can harm comfort.

Managers should ask whether air quality is measured or assumed. They should ask which spaces are monitored, how alerts are handled, how filters and ventilation equipment are maintained, and whether complaints about stuffiness, odor or sleep quality are linked to engineering data. The goal is not to overwhelm managers with technical metrics. The goal is to use measurement to prevent invisible discomfort from becoming visible dissatisfaction.

From Technical Failure to Financial Consequence

Technical failures become financial consequences through several pathways. The first is immediate compensation. If a room is too hot, has no hot water or suffers from noise, the hotel may offer a discount, refund, amenity, loyalty points or room move. The direct cost may be small for one guest but significant if repeated across many rooms.

The second pathway is room availability. A technical defect may take a room out of order. If the hotel is full or near full, every out-of-order room represents lost revenue. If several rooms are affected by a systemic issue, the revenue impact becomes measurable quickly. Out-of-order rooms also disrupt housekeeping, front office allocation and revenue management.

The third pathway is operational labor. Technical failures consume staff time. Front Office handles complaints, Housekeeping reports and revisits rooms, Engineering responds under pressure, managers coordinate solutions, and Finance may process compensation. The visible cost is not only repair; it is organizational distraction.

The fourth pathway is reputation and future demand. Negative reviews about comfort, water, noise, smell or safety can reduce conversion, affect group business and weaken brand trust. A guest who posts about a hot room or bad shower is not reviewing a technical system. They are reviewing the hotel.

The fifth pathway is emergency expenditure. Reactive repairs often cost more than planned interventions. Emergency contractors, urgent parts, overtime, compensation and operational disruption accumulate quickly. Preventive maintenance may appear expensive until compared with crisis response. A managerial mindset should therefore consider lifecycle cost, not only immediate cash outflow.

Managerial Decision-Making Under Limited CAPEX

Infrastructure decisions often occur under budget constraints. Ownership may push sustainability targets while limiting CAPEX. Energy costs may rise while guest complaints increase. Engineering may request investment in systems that are invisible to guests. Operations may request visible improvements. Finance may ask for short-term cost control. The manager must arbitrate.

A disciplined decision process begins with risk mapping. Which system failures would most severely affect guests, safety, compliance, revenue or asset value? Which failures are already happening? Which are likely? Which can be mitigated through maintenance, training or controls, and which require capital investment?

The second step is guest-impact analysis. Infrastructure investments should be evaluated based on how directly they affect comfort, reliability and trust. An HVAC upgrade that reduces complaints in many rooms may produce stronger guest value than a visible project with limited operational impact. On the other hand, a technical project with little guest or risk benefit may not be the first priority.

The third step is financial logic. Managers should distinguish CAPEX from OPEX, immediate repair from long-term replacement, energy saving from guest impact, and compliance necessity from discretionary improvement. Some investments pay back through energy savings. Others protect revenue by preventing failure. Others are required for safety or legal reasons regardless of payback. The recommendation to ownership must make this logic clear.

The fourth step is sequencing. Limited CAPEX does not mean doing nothing. It means phasing. The hotel may begin with monitoring, preventive maintenance and operational controls while preparing larger investment. It may prioritize high-risk zones first. It may combine staff training with technology. It may negotiate with owners using evidence: complaints, energy data, maintenance logs, out-of-order rooms and risk scenarios.

Communicating with Engineers, Contractors and Owners

A hospitality manager must communicate across technical and executive worlds. With engineers, the manager should be precise about symptoms, timing, location, guest impact and operational urgency. Vague statements such as “the room is bad” are less useful than “three rooms on the fifth floor reported weak hot water between 7:00 and 8:30, and two guests requested compensation.” Good operational reporting helps Engineering diagnose patterns.

With contractors, managers should clarify scope, response time, documentation, safety procedures, guest-impact mitigation and follow-up. Contractors may be highly skilled technically but unfamiliar with hotel service priorities. A repair that is acceptable in a closed office building may not be acceptable in an occupied luxury hotel. Noise, access, cleanliness, timing, communication and guest visibility matter.

With owners, managers should translate technical needs into business language. Owners need to understand risk, cost, options and consequences. Instead of saying "the HVAC is old," management should explain the complaint trend, affected rooms, energy inefficiency, maintenance cost, failure risk, guest compensation, investment options and recommended timeline. The more clearly the manager connects infrastructure to revenue, reputation and asset value, the more credible the recommendation becomes.

This translation skill is one of the most important learning outcomes of this chapter. Future managers must not hide behind technical language, but they must also not oversimplify. They must create a bridge between plant room reality and executive decision-making.

CHAPTER SUMMARY

Hotel infrastructure is the physical foundation of guest experience. Guests judge visible outcomes, but those outcomes depend on invisible systems that operate continuously. HVAC, plumbing, electrical systems, emergency power, heat, light and power distribution, IoT controls, waste management, sustainability practices and indoor air quality all influence comfort, safety, cost and reputation.

The chapter emphasized that infrastructure is not only technical. It is strategic. A failure in temperature, water, power, ventilation or drainage can become a guest complaint, a compensation issue, a revenue loss, an online review problem, a compliance exposure or an asset-value concern. Managers must therefore understand infrastructure as risk governance.

HVAC is central to comfort because it affects temperature, ventilation, humidity, air quality and noise. Plumbing and water management shape guest satisfaction through hot water, pressure, drainage and hygiene. Electrical systems and emergency power support safety, communication, access, lighting and operational continuity. Distribution and zoning determine whether energy is used intelligently or wasted. IoT can improve control, but only when connected to clear operational ownership. Waste and sustainability require daily systems, not slogans. CO2 monitoring links air quality, wellness and energy efficiency.

The managerial responsibility is not to become a technician. It is to ask the right questions, understand trade-offs, connect technical information to business impact, communicate clearly with engineers and owners, and make decisions under risk. The best managers use infrastructure knowledge to protect the guest, the staff, the asset, the owner and the brand.

OPERATIONAL CASE STUDY

The Venturo City Hotel Infrastructure Dilemma

The Venturo City Hotel is a 4-star, 180-room urban hotel located in a major European business district. The property performs well during weekdays due to corporate demand and meetings, but weekend leisure demand has become more competitive. The hotel was last renovated nine years ago. Public areas still look acceptable, but several technical systems are aging. Ownership has asked the General Manager to improve profitability, reduce energy consumption and protect online reputation without requesting a major renovation budget this year.

Over the past three months, the hotel has experienced a pattern of guest complaints. Some guests report that rooms are too warm at night, especially on upper floors. Others mention noisy air conditioning. Meeting clients have complained that conference rooms feel stuffy after one hour of use. Several reviews mention that the hotel "does not feel fresh." At the same time, utility bills have increased by 18% compared with the same period last year. The Chief Engineer explains that the HVAC system is functional but aging, that filters and controls need closer attention, and that zoning is not ideal. He also notes that the team is responding to complaints reactively because preventive maintenance has been compressed due to staffing constraints.

Water complaints have also increased. During peak morning periods, some rooms experience inconsistent hot-water temperature and weak pressure. The issue is not constant, which makes it difficult for Front Office to predict. Housekeeping reports occasional bathroom odors on two floors, but these are often solved temporarily and then reappear. Engineering suspects a combination of drainage trap issues, peak demand and aging pressure regulation.

A recent internal review also raised concerns about emergency power readiness. The backup generator is tested monthly, but the test reports are inconsistent. The hotel is not fully clear which systems remain operational during a prolonged power failure. The Director of Finance is reluctant to approve external testing because several other cost-saving initiatives are underway.

Ownership has offered limited CAPEX for the year. The hotel cannot fund a full HVAC replacement, plumbing upgrade and generator modernization at the same time. The Executive Committee must decide how to prioritize investment and operational action. The General Manager asks the leadership team to prepare a recommendation that balances guest satisfaction, safety, energy cost, sustainability, operational risk and owner expectations.

Case Questions

1. Which infrastructure issue represents the highest immediate risk, and why?
2. Which issues are primarily guest-experience risks, and which are safety or compliance risks?
3. What information should the General Manager request from Engineering before deciding on CAPEX priorities?
4. How should the hotel distinguish between problems that require immediate operational action and problems that require capital investment?
5. What low-CAPEX actions could be taken within 30 days to stabilize the situation?

6. How should the hotel communicate with ownership to justify infrastructure investment?
7. What data should be monitored over the next three months?
8. How can Front Office and Housekeeping improve the quality of technical reporting?
9. Where could IoT or smart monitoring add value without becoming an unnecessary gadget?
10. What would be your final recommendation to the Executive Committee?

Suggested Executive Analysis

A strong executive analysis should avoid choosing only the most visible complaint. The emergency power uncertainty may represent the highest safety and liability exposure, even if guests are not currently complaining about it. The HVAC and air-quality issues represent both guest satisfaction and energy-cost risks. The water-pressure and odor issues represent immediate comfort, hygiene perception and possible system deterioration. The correct managerial response is therefore not a single technical fix but a phased governance plan.

Within 30 days, the hotel should strengthen documentation and testing of emergency power, complete a targeted HVAC maintenance and filter program, map complaints by room and floor, inspect drainage traps and pressure issues on affected floors, and create a cross-department reporting protocol. It should also review meeting-room ventilation and consider temporary operational adjustments for high-occupancy events.

For CAPEX, the hotel should use evidence to build a prioritized proposal. If generator testing reveals compliance or life-safety gaps, this becomes non-negotiable. HVAC investment should be framed through guest complaints, energy cost, room impact and sustainability. Water-system interventions should be prioritized where peak demand and recurring complaints are proven. IoT sensors may be useful if they provide actionable data on CO2, occupancy, temperature and energy, but only if Engineering owns the response process.

The recommendation to ownership should be business-focused: protect life safety first, reduce recurring guest complaints second, and target energy savings through controls and maintenance while preparing larger phased investment. The hotel should not present infrastructure as a technical wish list. It should present it as revenue protection, margin improvement, risk mitigation and asset preservation.

Chapter Key Terms

- **Building infrastructure:** The physical systems that allow the hotel to operate, including HVAC, plumbing, electricity, power distribution, ventilation, waste systems and controls.
- **HVAC:** Heating, ventilation and air conditioning; the systems that regulate temperature, airflow, humidity and air quality.
- **Ventilation:** The process of supplying fresh air and removing stale air from indoor spaces.
- **Zoning:** The division of a building into areas that can be controlled separately for heating, cooling, lighting or power use.

- **Centralized HVAC:** A system where heating or cooling is produced centrally and distributed through the building.
- **Decentralized HVAC:** A system where heating or cooling is controlled locally, often by room or smaller zone.
- **Preventive maintenance:** Planned maintenance designed to reduce failure risk before equipment breaks down.
- **Emergency power:** Backup electrical supply used to maintain critical systems during a main power failure.
- **Backup generator:** Equipment that provides electricity to selected systems when normal power is interrupted.
- **Power distribution:** The way electrical power is delivered from supply sources to circuits, rooms, equipment and systems.
- **Building Management System:** A control platform used to monitor and manage building systems such as HVAC, lighting and energy use.
- **IoT:** Internet of Things; connected sensors and devices that collect data and support automatic or semi-automatic control.
- **CO2 monitoring:** Measurement of carbon dioxide levels as an indicator of ventilation adequacy and indoor air freshness.
- **Indoor air quality:** The condition of air inside the building, including freshness, ventilation, filtration, humidity and perceived comfort.
- **Legionella risk:** A water-system health risk that requires careful temperature, circulation, monitoring and maintenance practices.
- **CAPEX:** Capital expenditure; investment in assets or major improvements with long-term value.
- **OPEX:** Operating expenditure; recurring costs required for daily operations, maintenance and utilities.

- **ESG:** Environmental, social and governance criteria used to evaluate organizational responsibility and reporting.
- **Deferred maintenance:** Maintenance that is postponed, often creating greater risk or future cost.
- **Asset value:** The long-term economic value of the hotel property and its systems.

CHAPTER REVIEW QUESTIONS

- Why can hotel infrastructure be described as the physical foundation of guest experience?
- Explain how a technical HVAC issue can become a revenue and reputation problem.
- What is the managerial difference between understanding a system and being able to repair it?
- Compare centralized and decentralized HVAC from the perspective of comfort, cost and risk.
- Why is hot-water production both a guest-satisfaction issue and an energy-cost issue?
- Which hotel systems should remain operational during a main power failure, and why?
- How can poor zoning increase energy waste without improving guest comfort?
- What conditions must be present for IoT to create real operational value in a hotel?
- Why should sustainability be treated as daily operational management rather than marketing communication?
- In a limited CAPEX situation, what criteria should managers use to prioritize infrastructure investment?

CHAPTER REFERENCES

- Course presentation: Hotel Building Infrastructure & Technical Systems, Basics of Hospitality Engineering, Course 2.
- Teacher notes: C2 - Hotel Building Infrastructure & Technical Systems, Pedagogical Objective and Executive Teaching Notes.
- ASHRAE. ASHRAE Handbook series: HVAC Applications and HVAC Systems and Equipment. American Society of Heating, Refrigerating and Air-Conditioning Engineers.
- CIBSE. Guide M: Maintenance Engineering and Management. Chartered Institution of Building Services Engineers.
- CIBSE. Guide F: Energy Efficiency in Buildings. Chartered Institution of Building Services Engineers.
- ISO 50001. Energy management systems - Requirements with guidance for use. International Organization for Standardization.
- ISO 41001. Facility management - Management systems - Requirements with guidance for use. International Organization for Standardization.
- World Health Organization. Guidelines and technical resources on indoor air quality, ventilation and water safety in buildings.
- Health and Safety Executive. Guidance on Legionella and water system management in buildings.
- British Standards Institution. BS EN standards and guidance related to emergency lighting, electrical safety and building services management.

Chapter 3

Fire, Life Safety and Regulatory Constraints

Safety as a Managerial Responsibility

CHAPTER GENERAL OVERVIEW

Fire, life safety and regulatory constraints are often introduced to hospitality students as technical subjects: alarms, sprinklers, extinguishers, evacuation plans, inspection reports and compliance certificates. In hotel management, however, these topics must be understood first as leadership responsibilities. A hotel manager does not need to fight fires personally, design sprinkler systems or calculate detector placement. The manager must prevent leadership failure before an emergency occurs.

This chapter develops that executive perspective. Fire safety is not a back-of-house technical issue. It is a direct responsibility toward guests, employees, owners, insurers, regulators and the brand. Guests sleep in unfamiliar buildings. They trust the hotel to maintain detection systems, keep evacuation routes clear, train staff, test emergency lighting, protect fire doors, respond calmly and communicate correctly. If a fire alarm fails, if a sprinkler valve is not properly maintained, if a staircase is obstructed, or if staff members do not know who is in command, the failure is not only technical. It is managerial.

The previous chapters established that engineering is the foundation of business continuity and that infrastructure directly shapes guest experience. This chapter takes the next step: safety is executive accountability. Fire and life safety systems protect the ability of the hotel to operate, but more importantly they protect life. They also protect the hotel from legal exposure, insurance disputes, operational closure, reputational collapse and loss of owner confidence.

The guiding sentence for this chapter is: you do not delegate responsibility, only tasks. An engineer may test a system. A contractor may inspect equipment. A security officer may support evacuation. A department head may brief staff. Yet the hotel leadership remains accountable for ensuring that the system exists, is maintained, is documented, is funded, is understood and is operational when needed. A General Manager or senior leader cannot defend a failure by saying, "I did not know." In safety governance, not knowing may itself indicate a lack of oversight.

The chapter introduces the main elements of fire and life safety in hotels: the specific risk profile of hotels, detection systems, active suppression, passive fire protection, evacuation logic, crisis leadership, legal and insurance realities, maintenance obligations, documentation and reputation risk. The content remains managerial rather than engineering-mechanical. Students should understand enough to ask the right questions, challenge weak assumptions, request documentation, support preventive budgets and make decisions that protect life before they protect operations.

Managerial principle: fire safety is not optional, not purely technical and not delegable. It is a leadership responsibility that must be managed before the alarm sounds.

CHAPTER CONTENT

Fire Safety as Executive Responsibility

The first instinct in many organizations is to place fire safety entirely under Engineering or Security. This instinct is understandable but incomplete. Engineering teams maintain detection panels, pumps, valves, emergency lighting, generators and fire doors. Security teams may monitor alarms, assist with evacuation and coordinate with emergency services. These functions are essential. Yet neither function removes the responsibility of senior hotel leadership to govern the overall risk.

The executive responsibility begins with recognition. Fire safety is not a technical niche. It is the hotel's front line of defense against the most serious operational scenario: a threat to life in a building full of guests. The hotel is a legal entity with obligations. The General Manager and senior leaders are operationally accountable for ensuring that these obligations are taken seriously. Leadership must create the conditions for safety: budgets, inspections, staff training, authority lines, emergency plans, documentation and disciplined follow-up.

In a crisis, the difference between a controlled incident and a catastrophe often lies in the quality of preparation. A system that has not been tested may not function. A staff member who has never rehearsed evacuation may hesitate. A fire door that has been propped open for convenience may allow smoke to travel. A manager who has never clarified command authority may create confusion at the worst moment. Fire safety therefore starts long before a fire. It starts in leadership meetings, preventive maintenance planning, induction training, night-manager briefings and monthly review of compliance logs.

The phrase "No insurance policy covers poor leadership" captures an important truth. Insurance may cover some financial losses, depending on the policy and circumstances, but it cannot erase legal scrutiny, media attention, guest trauma, staff fear, damaged reputation or owner anger. It cannot restore a culture that treated safety as paperwork. Leaders must therefore approach fire safety as a moral, legal and commercial responsibility.

Why Hotels Are High-Risk Environments

Hotels are unique risk environments because they combine vulnerability and complexity. Unlike office buildings, many occupants are asleep during peak risk hours. Unlike residential buildings, many occupants are unfamiliar with the layout. Unlike small restaurants, hotels may contain guestrooms, kitchens, laundries, meeting spaces, bars, storage areas, plant rooms and back-of-house corridors within the same property. A hotel may also operate at high occupancy while hosting conferences, weddings or banquets, creating a dense population of guests who may not know the exits.

The 24/7 nature of hotel operations increases exposure. A technical issue can occur at 2:00 AM, when staffing levels are lower and guests are asleep. The night team may be smaller, less experienced or less supported by senior managers on-site. Yet night-time incidents often carry greater life-safety risk because evacuation is slower and disorientation is stronger. A hotel must therefore be designed and managed for the weakest operational moment, not only for the daytime leadership presence.

Guest unfamiliarity is a major factor. Guests do not know the building like employees do. They may not remember the nearest stairwell. They may not read the evacuation plan on the back of the guestroom door. They may panic when alarms sound. They may try to return to rooms for personal belongings. They may use elevators despite instructions. They may not understand announcements in the local language. These realities mean that evacuation plans must be simple, visible, rehearsed and supported by trained staff.

High-risk zones intensify the challenge. Kitchens involve heat, grease, gas, extraction systems and late-night preparation. Laundries involve heat, lint, electrical equipment and chemicals. Bars and events may involve alcohol, music, crowds and reduced attention to emergency cues. Guestrooms may include personal electrical devices, chargers, candles or other behaviors the hotel cannot fully control. Storage areas may accumulate combustible materials if not supervised. Engineering and management must therefore identify the full risk map of the property, not only the obvious public areas.

Fire Safety and Business Continuity

A fire incident is not only about flames. It is about everything that follows. Smoke can travel quickly and affect multiple floors. Panic can create injuries even before flames spread. Evacuation may be required. Guests may need relocation. Operations may stop immediately. Fire services, police, regulators, insurers and owners may all become involved. Media coverage can begin within hours, especially if guests post videos or comments online.

Business continuity depends on prevention and preparedness. A contained kitchen fire with functioning suppression, trained staff and clear command may result in limited disruption. The same fire with delayed detection, unclear staff response and poor documentation may result in evacuation, closure, investigation and severe reputational damage. The technical event may be similar; the management outcome may be radically different.

The long-term impact of a safety incident can exceed the immediate repair cost. A hotel may face room cancellations, event cancellations, insurance delays, legal costs, license review, staff anxiety, owner intervention and negative reviews. If the narrative becomes "hotel neglected safety," recovery can take years. In hospitality, reputation is not a soft issue. It is an asset linked to occupancy, ADR, talent attraction, owner confidence and brand credibility.

Managers must therefore view fire safety as a form of business continuity planning. The objective is not merely to comply with regulations. The objective is to ensure that life safety systems, staff behavior, communication protocols and leadership decisions are ready under pressure. Compliance is the minimum. Leadership maturity goes further: it asks whether the hotel is genuinely prepared.

GM and Chief Engineer: Different Responsibilities

A clear distinction must be made between technical execution and business accountability. The Chief Engineer focuses on equipment specifications, system maintenance, compliance testing, defect reports, contractor coordination and technical execution. The General Manager and executive team focus on business responsibility, operational accountability, strategic oversight and decision-making authority. Both roles are essential, but they are not interchangeable.

The engineer may report that a fire panel requires attention, a sprinkler valve needs correction, emergency lighting tests have failed, or fire doors are being blocked by operational practices. Once such information is known, the managerial question is no longer technical. It becomes a decision about risk. Will the hotel fund the repair? Will it close affected areas? Will it bring in external specialists? Will it train staff again? Will it accept temporary restrictions? Will it escalate to ownership?

The danger appears when leadership treats engineering warnings as operational inconvenience. Deferred maintenance may seem like cost control in the short term, but in life safety it can become deferred disaster. A manager who delays repairs, reduces training or ignores documentation weaknesses is effectively accepting risk on behalf of the hotel, the guests and the owner. That decision should never be unconscious.

The managerial rule is simple: engineers execute, managers decide. Managers do not need to know every technical detail, but they must know enough to understand the consequence of inaction. They must ask what happens if the system fails, who is exposed, what records exist, what regulations apply, what the insurer would request, and whether current controls are adequate. Leadership is not technical mastery; it is responsible oversight.

Fire Detection Systems

Fire detection systems are designed to provide early warning. If detection fails, every other response is delayed. In hotels, early warning is essential because guests may be asleep, staff may be dispersed and the building may be densely occupied. Detection systems create the first link in the chain of survival: identify the threat, alert people, activate response and support evacuation.

Common detection components include smoke detectors, heat detectors, manual call points, alarm panels and connections to monitoring services or emergency response systems where applicable. Smoke detectors identify particles in the air. Heat detectors respond to temperature thresholds or rapid temperature increase. Manual call points allow occupants or staff to trigger the alarm. Alarm panels centralize information, show zones and support diagnosis. Monitoring connections can alert external responders or a monitoring provider depending on local arrangements.

Managers do not need to select detector types, but they must understand detection as a governance issue. Are all areas covered according to applicable requirements? Are alarms tested? Are faults investigated quickly? Is the panel monitored? Do night staff understand what signals mean? Are false alarms reviewed seriously? Are repeated faults tolerated? Is the system connected to clear response procedures?

False alarms deserve particular attention. Some hotels become desensitized when alarms trigger too often. Staff begin to assume that the next alarm is also false. This is dangerous. A culture that treats alarms casually weakens response discipline. Management should track false alarms, identify causes, correct technical issues and reinforce that every alarm is treated seriously until verified by authorized procedures.

Fire Suppression Systems

Detection warns; suppression acts. Fire suppression systems are designed to control or extinguish fires, reduce spread and buy time for evacuation and emergency response. Hotels may use sprinklers, portable extinguishers, kitchen hood suppression systems, gas shut-off systems and other specialized solutions depending on building type, local regulation and operational risk.

Sprinkler systems are among the most important active fire protection tools. They are not designed to preserve décor or avoid inconvenience. They are designed to save lives and control fire spread. Aesthetic concerns must never override safety. If design teams, owners or operators resist visible safety features for visual reasons, the manager must defend the principle that life safety comes first.

Fire extinguishers are portable manual tools, but their value depends on correct type, correct placement, inspection and staff training. A staff member should not improvise with the wrong extinguisher or take personal risk beyond training. Kitchen hood suppression is particularly important because grease fires require specific systems. Gas shut-off systems can reduce escalation by cutting fuel supply. These systems require regular inspection and documentation.

The managerial focus is readiness. Are systems inspected? Are defects closed? Are extinguishers accessible and not hidden behind furniture or storage? Are kitchen suppression systems maintained by qualified providers? Are staff trained on when not to fight a fire? Are contractors' reports reviewed by management or simply filed? Suppression systems are only valuable if they work immediately and are supported by human discipline.

Passive Fire Protection

Passive fire protection is often less visible than alarms or sprinklers, but it is critical. Passive systems are built into the structure and layout of the building. They include fire doors, compartmentation, fire-resistant walls, protected stairwells, fire-resistant materials and stairwell pressurization where applicable. Their purpose is to contain fire and smoke, preserve evacuation routes and slow the spread of danger.

Passive systems can fail silently. A fire door that is propped open may look harmless during daily operations. It may even seem convenient for housekeeping, banquets or back-of-house movement. In a fire, that same door can allow smoke to spread quickly. A blocked stairwell may be unnoticed during normal operations but fatal during evacuation. Damaged seals, missing closers, poor compartmentation or unauthorized penetrations through fire-rated walls can weaken the building's protection without attracting daily attention.

This is why operational behavior matters as much as equipment. Fire doors must close. Corridors must remain clear. Storage must not accumulate in evacuation routes. Contractors must not compromise fire-rated barriers without proper restoration. Housekeeping, F&B, banqueting, security and engineering teams must understand that convenience cannot override life safety. Leaders must enforce this consistently, not only before inspections.

Managers should regularly walk the property with fire safety in mind. Are exits visible? Are evacuation routes clear? Are doors closing properly? Are signs illuminated? Are housekeeping trolleys or banquet equipment creating obstructions? Are back-of-house areas disciplined? The leadership walk is not a technical inspection. It is an accountability habit.

Evacuation Logic and Human Behavior

Evacuation is not a diagram; it is human behavior under stress. Guests may be tired, disoriented, intoxicated, elderly, disabled, with children, carrying luggage or unable to understand instructions. Panic can override logic. People may follow crowds rather than signs. Some may return to rooms. Some may use elevators. Some may freeze. Fire safety planning must therefore assume imperfect behavior and design systems and procedures that remain effective under pressure.

Clear signage and emergency lighting are essential because visibility may be reduced by smoke or power failure. Evacuation maps must be accurate and understandable. Exit routes must be intuitive. Assembly points must be defined. Staff must know where to guide guests. Special attention must be given to guests with reduced mobility and to areas with high noise, such as events or bars, where alarms or announcements may be harder to perceive.

Staff training is critical. In many emergencies, guests look to employees for cues. A calm, trained employee can reduce panic. An uncertain employee can spread confusion. Training should therefore include not only procedures but also behavior: how to speak, where to stand, when to escalate, what not to say, how to avoid speculation and how to prioritize life safety over operational convenience.

The principle of single command is important. During a crisis, multiple people giving conflicting instructions can create danger. The hotel must know who takes command, who communicates with emergency services, who manages guest information, who supports evacuation, who handles media or owner communication, and who documents decisions. These roles should be defined before the incident, not invented during it.

Crisis Communication and Managerial Decisions

When a fire alarm triggers, managers must make decisions quickly and calmly. The first priority is life safety. Operations, revenue, guest inconvenience and event disruption are secondary. A manager should never silence or reset an alarm casually to protect guest comfort or avoid embarrassment. Verification must follow approved procedures, and the protection of life must guide every decision.

The first managerial decisions typically concern command, verification, evacuation, communication and documentation. Who is in charge? What information is confirmed? Are emergency services contacted? Are guests being guided? Are staff following the plan? Are elevators being controlled? Are mobility-impaired guests supported? Is the incident log being maintained? Who communicates internally and externally?

Communication must be factual, calm and limited to verified information. Guests do not need technical speculation. They need clear instructions, reassurance and visible leadership. Staff should not provide contradictory explanations. The hotel should use predefined channels, scripts and escalation protocols. After the incident, communication with owners, insurers, regulators and possibly media must be aligned with legal guidance and accurate documentation.

Documentation begins during the event. Time of alarm, actions taken, decisions made, emergency services contact, evacuation status, system behavior, staff actions and guest issues should be recorded. This is not bureaucracy. It protects the hotel by creating an accountable record. In the absence of documentation, memories conflict and legal exposure increases.

Legal and Insurance Reality

The legal reality of fire safety is uncomfortable but necessary for future managers to understand. The hotel as a legal entity is liable for safety failures according to applicable laws, regulations, contracts and duties of care. The General Manager or senior leadership may be operationally accountable for oversight. A contractor may have technical responsibility for specific work, but the hotel cannot simply transfer its duty to maintain a safe environment.

“I did not know” is not a strong defense in leadership. If a system was not tested, if training was not completed, if maintenance was deferred, if reports were ignored, or if documentation cannot be produced, the issue becomes one of governance. Managers must be able to show that they had a system of oversight, that issues were escalated, that decisions were documented, and that reasonable action was taken.

Insurance investigations are evidence-based. Insurers may request maintenance logs, training records, compliance certificates, inspection reports, contractor reports, incident response plans and decision logs. If the hotel cannot prove that systems were maintained and staff trained, coverage may be challenged, delayed or reduced depending on the policy and circumstances. The practical rule is clear: if it is not logged, it may be treated as if it did not happen.

This is why documentation should be treated as part of safety management, not administrative decoration. A maintenance log is evidence of discipline. A training record is evidence of preparation. A corrective-action tracker is evidence of management follow-up. A signed inspection report is evidence of professional oversight. Good documentation does not replace good practice, but it proves that good practice exists.

Maintenance Obligations and Preventive Discipline

Life safety systems require regular maintenance. Detection systems, sprinkler systems, emergency lighting, fire doors, extinguishers, kitchen suppression systems, alarms, signage and emergency power must be inspected and tested according to local regulation, manufacturer guidance, insurance requirements and professional standards. The exact frequency varies by jurisdiction and system, but the leadership principle is constant: schedule it, log it and enforce it.

The teaching materials highlight examples such as monthly alarm checks, quarterly sensor testing, sprinkler inspections, emergency-lighting battery tests and fire-door checks. These examples show the rhythm of preventive discipline. A hotel cannot wait for annual inspection to discover that daily operations have weakened safety. Small checks, repeated consistently, protect the system.

Deferred maintenance in life safety is particularly dangerous because failure may remain hidden until the emergency. A guest may not know that a fire door closer is defective. A manager may not notice that a sprinkler valve has an issue. A staff member may not realize that emergency lighting batteries are weak. The system seems acceptable until the moment it is needed. This is why preventive maintenance must be funded even when no visible problem exists.

Managers must resist the temptation to treat safety maintenance as a flexible cost. During budget pressure, it may be tempting to postpone tests, delay repairs or reduce training hours. This creates invisible risk. A mature manager understands that preventive spending is not waste. It is the cost of keeping the hotel legally, morally and operationally viable.

Reputation Risk and Brand Survival

Engineering failure can become reputation failure within hours. A fire incident involving poor safety management produces a powerful media narrative: guests were put at risk because the hotel neglected safety. Once such a narrative forms, technical explanations may not repair trust. The public does not easily distinguish between a valve issue, a maintenance delay, a training gap and a leadership failure. It sees a hotel that did not protect people.

The reputation timeline can be severe. On day one, the incident occurs. On day two, media and social posts begin. Within a week, bookings may fall and corporate clients may reconsider events. Within a month, staff morale may suffer and recruitment may become harder. Over the following year, recovery may require public relations, audits, investment, leadership changes and renewed owner confidence. The financial loss may be temporary, but the reputational wound may last.

For independent hotels, the impact can be especially serious because they may not have the crisis infrastructure of a large chain. For branded hotels, the damage can extend beyond one property to the brand promise. For luxury or upscale properties, safety failures contradict the expectation of care, discretion and excellence. Guests may forgive a small operational inconvenience; they are much less likely to forgive a perceived disregard for safety.

Reputation protection begins with operational truth. The best crisis communication cannot compensate for poor preparation. If logs are missing, staff are untrained and warnings were ignored, communication becomes defensive. If systems were maintained, staff acted correctly and leadership documented decisions, the hotel can communicate with credibility. Reputation is protected before the incident, not after it.

Fire Safety as Leadership Maturity

The central lesson of this chapter is that fire safety reveals leadership maturity. A technically complex system can only protect guests if management creates a culture of seriousness around it. This culture is visible in budget decisions, inspection follow-up, training attendance, property walks, contractor management, documentation discipline and the willingness to inconvenience operations in order to protect life.

Mature leaders do not wait for inspectors to impose discipline. They ask preventive questions. They invite uncomfortable reports from engineers. They track unresolved risks. They communicate with owners about necessary investments. They train night teams. They define command structures. They make sure that every department understands its role. They treat life safety as a shared responsibility, but not as a responsibility that can be diluted.

This maturity is especially important because safety work is often invisible when successful. A well-maintained fire door does not generate a review. A tested alarm panel does not increase ADR directly. A staff evacuation drill may feel inconvenient. Yet these actions protect the hotel from catastrophic failure. The manager must therefore defend the value of prevention even when its success is measured by the absence of disaster.

In future roles, students may become department heads, operations managers, asset managers, owners' representatives or General Managers. They may not be the person testing the alarm, but they may be the person approving budgets, reading reports, deciding whether to close a floor, responding to an insurer or briefing an owner after an incident. This chapter prepares them for that responsibility.

Managerial Questions for Fire Safety Oversight

A manager does not need to inspect every detector personally, but a manager must create a disciplined questioning rhythm. Fire safety governance begins with the quality of questions leaders ask before a crisis. These questions should be asked regularly, not only after an incident or before an external inspection. They help translate technical maintenance into executive control.

The first set of questions concerns system readiness. Are all detection zones operational? Are there repeated faults on the alarm panel? How quickly are defects closed? Are sprinkler valves inspected and accessible? Are extinguishers visible, appropriate and within inspection date? Are kitchen suppression systems maintained by qualified specialists? Are emergency lights tested under realistic conditions? Are fire doors closing properly? Are stairwells clear at night, during events and during housekeeping peaks?

The second set concerns human readiness. Which employees know what to do when the alarm sounds? Does the night team understand the same procedure as the day team? Are temporary staff, interns, banquet extras and subcontractors briefed? Who takes command if the GM is absent? Who communicates with guests? Who contacts emergency services? Who checks assembly areas? Who supports guests with reduced mobility? A safety plan that is understood only by senior managers is not a safety plan; it is a document.

The third set concerns evidence. Can the hotel produce maintenance logs within one hour? Can it show training records? Are inspection reports signed, dated and followed by corrective actions? Are unresolved defects visible to senior leadership? Are decisions to defer works documented with risk evaluation and owner awareness? Documentation should not be scattered across email inboxes, contractor folders and informal notes. It should be structured enough to support decision-making and defensible enough to withstand investigation.

The fourth set concerns investment. Which life-safety items require urgent funding? Which are compliance-critical? Which are risk-reducing but not yet mandatory? Which items have been postponed more than once? Which owner decisions could expose the hotel operationally? Managers should avoid presenting safety investment as a technical preference. It should be presented as risk mitigation, legal protection, brand protection and duty of care.

Executive question: if an incident happened tonight, could the hotel prove that systems were maintained, staff were trained and leadership acted on known risks?

Building a Fire Safety Governance Rhythm

Fire safety cannot depend on occasional attention. It requires a governance rhythm: a repeated management pattern through which risks are reviewed, actions are tracked and accountability remains visible. This rhythm connects Engineering, Security, Operations, Human Resources, Finance, ownership and external contractors. It turns safety from a folder of certificates into a living management process.

A practical governance rhythm may include a monthly safety review chaired by the GM or Hotel Manager, with the Chief Engineer, Security Manager, Rooms Division, F&B, Housekeeping, HR and Finance represented. The meeting should review open defects, upcoming inspections, training completion, drill outcomes, contractor recommendations, recurring false alarms, blocked-route observations and any operational behavior that creates fire risk. The purpose is not to blame departments. The purpose is to make risk visible and to ensure that safety is not lost between departments.

A weekly property walk can complement formal meetings. Senior managers should walk guest corridors, back-of-house areas, event spaces, kitchens, storage zones, technical access points and stairwells with a safety lens. They should look for blocked exits, wedged doors, missing signage, poor lighting, combustible storage, housekeeping obstructions, overloaded sockets, unmanaged cables and informal workarounds. These walks communicate cultural seriousness. Staff understand what leadership truly values by what leadership checks.

Training should also follow a rhythm. New employees should receive life-safety induction. Departmental refreshers should occur regularly. Night-team scenarios should be rehearsed. Banquet and event teams should practice crowd management. Front Office should practice guest communication. Housekeeping should understand fire doors and corridor obstructions. F&B should understand kitchen suppression, grease risk and gas shut-off logic. Security and Engineering should rehearse command coordination. A drill is not a theatrical exercise; it is a rehearsal for human behavior under pressure.

The governance rhythm must include ownership communication. Owners sometimes see safety investment as invisible cost, especially when no incident has occurred. Managers must educate owners that prevention protects the asset. A clear owner briefing should distinguish mandatory compliance, urgent corrective actions, risk-reduction investments and long-term resilience projects. When leaders explain safety in financial and reputational language, they make it easier for owners to support responsible decisions.

Departmental Implications of Fire and Life Safety

Fire safety is coordinated by Engineering and Security in many hotels, but every department contributes to success or failure. Front Office is often the communication point during an incident. It may receive calls from guests, coordinate rooming lists, support emergency services with occupancy information and help account for guests. If

Front Office is not trained, the hotel can lose control of information quickly.

Housekeeping has daily influence over passive fire protection. Room attendants, supervisors and public-area cleaners move through corridors, service areas and guestrooms constantly. They notice blocked doors, damaged closers, missing signage, suspicious smells, overloaded sockets or guest behaviors that create risk. They also may unintentionally create risk if trolleys block corridors or service doors are held open for convenience. Housekeeping must therefore be part of safety culture, not treated as an observer.

Food and Beverage is connected to some of the highest-risk zones in the property. Kitchens involve open flames or high heat, grease accumulation, extraction systems, gas lines, electrical equipment, cleaning chemicals and intense operational pressure. Banqueting adds temporary layouts, decorations, cables, furniture movement and large crowds. F&B managers must understand that commercial urgency cannot override safe layouts, clear exits and disciplined kitchen procedures.

Human Resources supports safety by ensuring induction, training attendance, records, refresher programs and accountability. Finance supports safety by ensuring that preventive maintenance and compliance expenses are budgeted realistically. Sales and Events support safety by not promising setups that compromise evacuation routes. The GM integrates these departmental responsibilities into one leadership system. Fire safety fails when everyone assumes someone else owns the detail.

From Compliance to Safety Culture

Compliance is essential, but compliance alone is not enough. A hotel may pass an inspection and still have weak daily habits. A certificate may be valid while staff remain confused. A system may be installed while management fails to act on faults. Safety culture means that people behave safely when no inspector is present, no crisis is happening and no senior leader is watching.

A strong safety culture is visible in small decisions. Staff close fire doors without being asked. Managers challenge blocked exits immediately. Department heads attend drills seriously. Contractors are supervised when their work affects fire-rated walls or ceilings. Defects are reported early. False alarms are investigated rather than dismissed. Budgets include preventive work. Senior leaders ask about safety with the same seriousness as revenue, guest satisfaction and payroll.

Creating this culture requires repetition. One memo is not enough. One training session is not enough. One inspection is not enough. Leadership must repeat the message until it becomes operational habit: life safety comes before convenience. This message may be uncomfortable because it sometimes slows down operations, increases cost or challenges aesthetic preferences. Yet that discomfort is the price of responsible hospitality management.

Future hotel leaders should remember that guests rarely reward safety directly because they assume it exists. That assumption is exactly why the responsibility is so serious. The absence of disaster is not luck when systems are governed well. It is the result of disciplined, invisible leadership.

CHAPTER SUMMARY

Fire, life safety and regulatory constraints must be understood as leadership and risk-management responsibilities, not as purely technical engineering tasks. Hotels are high-risk environments because they operate continuously, host sleeping guests, contain unfamiliar layouts, include high-risk zones and may concentrate large numbers of people during events. These characteristics make prevention, detection, evacuation planning, staff training and documentation essential.

Detection systems provide early warning through smoke detectors, heat detectors, manual call points, alarm panels and monitoring arrangements. Suppression systems such as sprinklers, extinguishers, kitchen hood suppression and gas shut-offs control fires and reduce escalation. Passive fire protection, including fire doors, compartmentation, protected stairwells and fire-resistant materials, slows the spread of fire and smoke. These systems must not only exist; they must be maintained, tested and respected by daily operations.

Evacuation is a human behavior challenge. Guests do not know the building and may panic under stress. Staff training, clear signage, emergency lighting, command structure and calm communication are critical. In an emergency, life safety takes priority over operations, revenue or inconvenience. Single command, factual communication and immediate documentation protect people and support legal and insurance review.

Legal and insurance realities make documentation central. The hotel is legally exposed when safety fails, and leadership is operationally accountable for oversight. Insurers and regulators may request maintenance logs, training records, inspection reports, compliance certificates and incident documentation. If it is not documented, it may not be defensible.

The strongest managerial lesson is that responsibility cannot be outsourced. Engineers and contractors execute technical tasks, but leaders decide priorities, approve resources and govern risk. Fire safety is not optional, not merely technical and not delegable. It is a measure of leadership maturity.

OPERATIONAL CASE STUDY

Fire Safety Failure During High Occupancy

The Hôtel Centralis is a four-star independent city-center property with 220 rooms across eight floors. It includes a full-service restaurant, three conference rooms, a ballroom, underground technical areas and a small rooftop bar. The hotel is not part of a chain, and although it has a competent Chief Engineer, the property has faced budget pressure for two consecutive years. Several preventive maintenance actions have been postponed because ownership has prioritized visible refurbishment projects and short-term cash control.

On a Saturday evening, the hotel is operating at 98% occupancy. A corporate conference of 120 attendees is using two meeting rooms, and a wedding reception of 80 guests is taking place in the ballroom. Many guests are staying overnight. The restaurant kitchen remains active late into the evening because of banquet service and late-room-service demand. The night team includes a duty manager, two front office agents, one security officer, a night auditor and limited engineering coverage on call.

At 11:47 PM, a grease fire starts in the kitchen during late-night preparation. The kitchen hood suppression system activates later than expected. At 11:49 PM, the fire alarm partially triggers, but only three of the eight floors receive the alarm signal. At 11:51 PM, sprinklers activate after a delay linked to a valve issue previously mentioned in a contractor report. At 11:53 PM, staff members begin giving conflicting instructions: one employee tells guests to remain calm in the lobby, another sends guests toward a stairwell, and a banquet supervisor attempts to keep wedding guests in the ballroom while waiting for confirmation.

The incident is eventually controlled by emergency services. No fatalities occur, but several guests are treated for smoke inhalation and anxiety. Videos appear online during the night showing confused guests in corridors and staff members disagreeing over instructions. By morning, local media report that the hotel's alarm system did not fully activate. The owner, insurer, legal counsel and local authorities all request documentation. The Chief Engineer states that he had previously flagged deferred maintenance, incomplete staff training and the absence of a formal night-command structure.

Managerial Analysis

The top managerial failures are not limited to the immediate technical malfunction. The first failure is deferred maintenance. A valve issue had already been identified, but leadership did not close the corrective action. The second failure is incomplete training. Staff members did not share a common evacuation logic, and their conflicting instructions increased confusion. The third failure is the absence of a formal command structure, especially for night operations. The hotel did not clearly define who had authority to decide, communicate and coordinate during a crisis.

Prevention would have required disciplined follow-up before the incident. The hotel should have maintained a corrective-action tracker for all life-safety defects, escalated unresolved safety items to ownership, funded urgent repairs, verified contractor recommendations and recorded decisions. Staff training should have included night-shift scenarios, role assignments, communication scripts, assembly-point management and clear instructions on when to evacuate. The command structure should have been documented and rehearsed, including the role of the duty manager, security officer, front office team, engineering support and emergency services liaison.

The next-morning executive briefing should be factual, structured and transparent. To the owner, the GM should explain the sequence of events, immediate guest safety actions, operational impact, known system failures, documentation status and urgent corrective measures. To the insurer, the hotel should provide maintenance logs, inspection reports, training records, contractor reports, incident timeline and emergency response documentation. To the media, communication should avoid technical speculation and focus on guest safety, cooperation with authorities, care for affected guests and immediate review of systems. Legal counsel should validate external messaging.

The lesson is that the incident was not created by fire alone. It was created by a combination of deferred maintenance, weak governance, incomplete training and unclear command. The engineer flagged issues. Leadership did not act with sufficient urgency. In fire safety, this distinction matters: systems do not save lives by themselves. Decisions, preparation and accountability make systems effective.

CHAPTER KEY TERMS

- **Active fire protection —** Systems or tools that detect, control or suppress fire through an active response, such as alarms, sprinklers, extinguishers, kitchen hood suppression or gas shut-off systems.
- **Alarm panel —** The central control and information point for a fire detection system. It indicates zones, faults, alarms and system status, helping trained staff and responders understand what is happening.

- **Business continuity** — The ability of the hotel to continue or restore operations after a disruptive event. Fire safety supports business continuity by reducing the scale and duration of incidents.
- **Command structure** — The predefined authority system that clarifies who makes decisions, communicates and coordinates actions during a crisis.
- **Compartmentation** — The division of a building into fire-resistant zones to slow the spread of fire and smoke and support safe evacuation.
- **Corrective action tracker** — A management tool used to record safety defects, responsible persons, deadlines, completion status and evidence of closure.
- **Emergency lighting** — Lighting designed to function during power failure or emergencies so that occupants can see evacuation routes and exits.
- **Evacuation logic** — The practical reasoning behind how people move safely out of a building during crisis, considering signage, routes, staff guidance, panic and guest unfamiliarity.
- **Fire detection system** — A system that identifies possible fire conditions through smoke detectors, heat detectors, manual call points, alarm panels and monitoring arrangements.
- **Fire door** — A rated door designed to resist fire and smoke for a specified period, helping to protect evacuation routes and contain danger.
- **Fire suppression system** — A system designed to control, reduce or extinguish fire, such as sprinklers, extinguishers or kitchen hood suppression.
- **Life safety** — The set of systems, procedures and management practices intended to protect people from harm during emergencies.

- **Maintenance log** — A written or digital record proving that inspections, tests, repairs and follow-up actions were completed.
- **Operational accountability** — The responsibility of managers to ensure that systems, teams and procedures are functioning effectively, even when technical tasks are delegated.
- **Passive fire protection** — Building features that slow fire or smoke spread without requiring activation, including fire doors, fire-resistant walls and protected stairwells.
- **Regulatory compliance** — The requirement to meet applicable laws, codes, standards, inspections and documentation obligations.
- **Reputation risk** — The possibility that a safety incident damages public trust, bookings, staff morale, owner confidence and brand value.
- **Single command** — The principle that one clearly identified authority coordinates decisions during an incident to avoid conflicting instructions.
- **Sprinkler system** — An automatic fire suppression system that releases water when heat activates sprinkler heads in affected areas.
- **Training record** — Evidence that employees have been trained on safety procedures, evacuation roles, communication protocols and emergency response expectations.

CHAPTER REVIEW QUESTIONS

- Why should fire safety in hotels be treated as a leadership responsibility rather than only as an engineering responsibility?
- What characteristics make hotels particularly vulnerable during fire or life-safety incidents?
- Explain the difference between the Chief Engineer's technical responsibilities and the General Manager's operational accountability.
- Why is early fire detection critical in a hotel environment, especially during night operations?
- How do active fire protection systems and passive fire protection systems complement each other?
- Why can a propped-open fire door be described as a leadership failure rather than only an operational inconvenience?
- What are the first managerial priorities when a fire alarm triggers during high occupancy?
- Why is documentation essential for insurance, legal review and management protection after a safety incident?
- How can deferred maintenance become a reputational risk for a hotel?
- In the operational case study, what were the three most important management failures, and what should leadership have done before the incident?

CHAPTER REFERENCES

- British Standards Institution. Fire detection and fire alarm systems for buildings. Consult the most recent applicable standards and local adaptations.
- European Union and national authorities. Fire safety, evacuation, workplace safety and public-access building regulations. Requirements vary by jurisdiction and must be verified locally.
- International Code Council. International Fire Code and related building-safety guidance. Consult the latest applicable edition and local amendments.
- National Fire Protection Association. Life Safety Code, Fire Alarm and Signaling Code, and fire protection guidance. Consult the latest applicable editions and local legal requirements.
- Occupational safety authorities and national labor agencies. Guidance on emergency preparedness, employee training, evacuation planning and workplace safety.
- Hotel brand standards and insurer guidance documents. Property-specific fire, life safety, emergency response, maintenance and documentation requirements.
- Course materials: Fire, Life Safety and Regulatory Constraints presentation and teacher notes prepared for Basics of Hospitality Engineering, Chapter 3.

Chapter 4 — Engineering Vocabulary and Operational Communication

Speaking Engineering as a Manager

CHAPTER GENERAL OVERVIEW

Hotel engineering is often misunderstood because managers hear technical problems before they understand business consequences. A chiller, a pump, an air handling unit, an alarm panel, or a BMS dashboard may sound like engineering language, but the managerial meaning is different. Behind each technical word sits a decision about guest comfort, operational continuity, safety, cost exposure, brand reputation, and asset value.

This chapter develops the ability to speak engineering as a manager. The objective is not to train hospitality students to repair equipment, read wiring diagrams, or choose technical specifications. The objective is to help future managers ask better questions, recognize risk, prioritize interventions, defend budgets, and translate operational incidents into executive decisions.

The chapter follows the logic of the course presentation: engineering is not a collection of tools; it is operational reliability. The teacher guidance emphasizes that most conflicts between Operations and Engineering are not caused by incompetence, but by language gaps. Engineering speaks systems. Management speaks impact. The manager's role is to connect both languages.

For a future General Manager, Operations Director, Asset Manager, or department head, this communication skill is critical. Owners rarely approve investment because a technical component has a problem. They approve investment when the issue is framed as risk reduction, guest experience protection, cost control, safety compliance, or asset preservation. A manager who cannot translate engineering language into business language becomes dependent on technical specialists. A manager who can translate becomes credible in front of the Executive Committee and ownership.

> *Good managers do not repair engineering problems. They translate problems into decisions.*

By the end of this chapter, students should understand the difference between corrective, preventive, predictive, planned, and reactive interventions. They should be able to identify whether an issue is urgent or important, whether it is a guest-impacting incident or a hidden asset risk, and whether a decision is required immediately or can be scheduled. They should also be able to prepare a concise executive engineering briefing with context, risk, options, recommendation, and decision required.

CHAPTER CONTENT

Engineering Communication as a Managerial Discipline

In many hotels, engineering is physically located behind the scenes. Technical rooms, maintenance workshops, plant rooms, storage areas, and service corridors are usually invisible to guests and often invisible to managers until something fails. This invisibility creates a managerial problem: engineering may be treated as a back-office function rather than as a strategic operating discipline.

Communication is the first step toward changing this perception. When Engineering reports "the chiller efficiency dropped by 18%," the message may sound technical. A manager may not immediately understand whether this is serious, whether guests are affected, whether the hotel must spend money, or whether the issue can wait. But translated into business language, the issue becomes clearer: the hotel is exposed to guest discomfort, rising energy cost, equipment shutdown risk, and potential room revenue loss.

A hospitality manager does not need to understand the internal mechanics of the chiller. The manager needs to understand what happens if the issue is ignored, who is impacted, how much risk is being accepted, and what decision is required. This is the central difference between technical knowledge and managerial literacy.

Engineering communication becomes especially important in moments of pressure. During an outage, leak, temperature complaint, kitchen ventilation issue, elevator failure, or safety concern, the quality of managerial questions determines the quality of the response. A manager who asks only "How long will it take?" may miss deeper issues. A stronger manager asks:

- What is the immediate guest impact?
- How many rooms, outlets, or operating areas are affected?
- Is there any safety, legal, or compliance exposure?
- Is this an isolated incident or a recurring pattern?
- What is the cost of temporary repair versus permanent correction?
- What decision is needed from management now?

These questions move the conversation from repair to decision-making. They also help Engineering communicate more effectively, because technical teams often know the risk but may not frame it in language that senior managers, owners, or finance teams can act upon.

Why Poor Engineering Language Creates Bad Decisions

Poor communication does not simply create confusion. It creates bad decisions. When technical issues are described without context, managers may underestimate risk. When everything is described as urgent, managers lose the ability to prioritize. When reports are filled with acronyms and technical readings, owners may delay investment because they cannot see the business case. When Engineering does not explain the cost of inaction, Finance may treat preventive maintenance as an expense to be reduced rather than as protection against future crisis.

A common operational conflict illustrates this point. Engineering may request budget for preventive maintenance on a boiler, kitchen hood, generator, or air handling unit. Operations may see no immediate problem and ask why money should be spent when the system still works. Finance may delay the cost until the next quarter. The Chief Engineer may insist that failure is likely, but without a clear business translation, the warning can be dismissed as technical anxiety.

The problem is not always the technical recommendation. The problem is often the framing. “The boiler needs servicing” is weaker than “If the boiler is not serviced before peak occupancy, the hotel risks hot water interruptions affecting guest rooms during the morning period, with potential compensation, complaints, and reputational damage.” The second statement gives management a reason to act.

In this sense, vocabulary is not decoration. Vocabulary creates control. A manager who understands the difference between corrective and preventive maintenance can challenge a budget proposal intelligently. A manager who understands downtime can ask about room availability and revenue impact. A manager who understands criticality can distinguish between a minor repair and a life-safety issue. Language gives credibility.

The Manager's Role: Not Technical Repair, but Executive Translation

One of the most important lessons in this chapter is to define clearly what managers are not expected to do. They are not expected to repair the equipment themselves. They are not expected to read complex wiring diagrams. They are not expected to select every technical brand or specification. In a professional hotel, those responsibilities belong to qualified engineers, specialists, contractors, consultants, and certified service providers.

However, managers are expected to lead the consequences of technical issues. They must prioritize interventions, allocate resources, defend budgets, monitor risk, and explain decisions to the Executive Committee, ownership, or asset management. They must also challenge technical teams when communication is unclear. Asking for clarity is not a sign of weakness; it is a managerial obligation.

The strongest managers act as translators between technical reality and business responsibility. They do not simplify engineering in a careless way. Instead, they translate complexity into decision logic. They ask Engineering to express problems in terms of impact, probability, urgency, and cost. They ask Finance to understand the operational risk of underinvestment. They help Operations understand why a planned maintenance shutdown may protect the guest experience in the long term.

This translation role becomes especially important in owner relations. Owners and asset managers think in terms of asset value, capital expenditure, risk, return, compliance, and reputation. They are less interested in the technical details of a pump than in the consequences of pump failure. A manager who can present both the risk and the decision clearly becomes a credible interlocutor.

Core Maintenance Vocabulary

The first vocabulary set every hospitality manager must master concerns maintenance logic. Maintenance is not one single activity. It includes several different approaches with different financial, operational, and risk implications. Confusing these categories leads to poor planning and weak budget decisions.

<u>Corrective Maintenance</u>

Corrective maintenance refers to action taken after a failure has occurred. It is the repair response to something that is already broken, malfunctioning, or no longer performing as required. Examples include repairing a guestroom air-conditioning unit after a complaint, fixing an elevator after breakdown, restoring hot water after an outage, replacing a failed pump, or responding to a kitchen equipment failure during service.

Corrective maintenance is sometimes unavoidable. Equipment can fail even in well-managed hotels. However, when a hotel is dominated by corrective maintenance, it usually reveals a deeper weakness. The engineering team is constantly reacting, guests are more likely to experience disruption, costs are harder to control, and staff work under stress.

From a managerial perspective, corrective maintenance often has four consequences. First, it is visible to guests because the failure has already occurred. Second, it creates pressure because the hotel must respond quickly. Third, it is frequently more expensive than planned intervention because emergency work involves urgency, overtime, temporary solutions, or specialist callouts. Fourth, it can damage reputation because guests remember discomfort more easily than they notice prevention.

Preventive Maintenance

Preventive maintenance refers to planned actions taken to avoid failure before it happens. Examples include replacing HVAC filters, inspecting pumps, servicing boilers, testing emergency lighting, cleaning kitchen extraction systems, verifying fire doors, checking water systems, maintaining elevators, and monitoring equipment cycles.

Preventive maintenance is often misunderstood because it looks like a cost when nothing is visibly wrong. This is why managers must be able to defend it. The value of prevention is not always visible in the month it is performed. Its value appears in the breakdowns that do not happen, the rooms that remain sellable, the guests who do not complain, the inspections that are passed, and the emergency repairs that are avoided.

The managerial challenge is that preventive maintenance requires discipline. It must be scheduled, budgeted, documented, and protected from short-term cost cutting. If every preventive task is postponed because the hotel is busy or because the budget is tight, the hotel gradually builds hidden risk. Emergency work is often not bad luck; it is delayed prevention.

Predictive and Planned Maintenance

Predictive maintenance uses data, readings, monitoring, patterns, or technology to anticipate failure before it becomes visible. For example, abnormal vibration in equipment, declining energy performance, recurring temperature variation, pressure instability, or unusual operating cycles may indicate future failure. Predictive maintenance is increasingly supported by building management systems, sensors, and digital monitoring tools.

Planned maintenance refers to interventions scheduled in a controlled way. The goal is to perform work at the least disruptive moment, with available parts, clear communication, and budget alignment. A planned intervention may still be corrective in purpose if it addresses a known problem, but it is managed proactively rather than as an emergency.

For managers, the distinction between reactive and planned work is strategic. Reactive work is unpredictable, visible, expensive, and stressful. Planned work is scheduled, controlled, budgeted, and usually invisible to guests. Strong hotels continuously try to move from reactive response to planned stability.

Maintenance Type	**Managerial Meaning**	**Typical Impact**
Corrective maintenance	Action after a failure has already occurred.	High urgency, guest disruption, often higher cost.
Preventive maintenance	Planned action to reduce the probability of failure.	Lower disruption, controlled budget, better asset protection.
Predictive maintenance	Use of data or monitoring to anticipate failure.	Better forecasting, targeted interventions, reduced surprise.
Planned intervention	Work scheduled at a controlled time with clear preparation.	Reduced operational impact and better communication.
Reactive intervention	Emergency response after an unplanned event.	Stress, cost escalation, reputational risk.

From Technical Issue to Business Impact

A central competency in engineering communication is the ability to translate a technical issue into business impact. This does not mean exaggerating the problem. It means making the consequences visible to decision-makers.

Consider the following technical statement: "The chiller efficiency dropped by 18%." For a non-technical manager, this may not immediately signal urgency. Translated into business language, it could mean that the hotel is using more energy to produce the same level of cooling, guestroom comfort may decline during peak temperatures, operating cost may rise, and the system may be closer to failure. The same technical statement now becomes a management issue.

Another example: "The kitchen hood inspection is overdue." This is not only a maintenance detail. It may represent fire risk, insurance exposure, regulatory non-compliance, operational interruption, and potential closure of a revenue-generating outlet. If the hotel has a full restaurant, banqueting operation, or conference business, the business impact can be significant.

A third example: "The boiler pressure is unstable." This may sound like a technical reading, but the managerial meaning is much broader. The hotel may face hot water inconsistency, guest complaints during morning peak periods, safety risks, energy inefficiency, and the possibility of emergency shutdown. The appropriate managerial response is not to ask which valve is involved, but to ask about risk level, guest exposure, timing, cost, and recommended decision.

The Translation Framework

Every engineering issue should be translated through a simple decision framework. The framework contains four essential questions.

- What is the issue? The problem must be expressed clearly and without unnecessary jargon.
- What happens if we do nothing? The risk of inaction must be visible.
- Who is impacted? The answer may include guests, staff, operations, finance, ownership, insurers, or regulators.
- What decision is needed? A report without a decision request has limited managerial value.

This framework forces clarity. It prevents technical issues from remaining isolated within Engineering. It also prevents managers from receiving long lists of problems without understanding priority. The final question is especially important: what decision is needed? If no decision is requested, the communication may create awareness but not action.

A well-translated engineering issue should allow an executive to decide. For example: "The main kitchen extraction system requires urgent cleaning and inspection. If delayed, the hotel faces increased fire risk, insurance exposure, and possible disruption to restaurant operations. The impacted areas are the main kitchen, banqueting production, and evening service. The recommended decision is to approve an overnight specialist intervention this week at an estimated cost of X, with minimal guest disruption."

This form of communication is clear, business-oriented, and decision-ready. It does not hide the technical reality, but it makes the managerial stakes visible.

Engineering Reporting for Managers

Engineering reports are only useful if they support decision-making. A weak report may contain many technical details but no prioritization. A strong report tells management what matters, why it matters, what could happen, what action is recommended, and by when.

Good engineering reports typically include a short description of the issue, a risk level, the cost of inaction, the recommended action, and the timing urgency. They avoid unnecessary jargon and explain the relationship between the technical issue and guest experience, safety, revenue, cost, compliance, or asset value.

Bad reports create several problems. Technical lists without context may overwhelm managers. Acronyms may reduce clarity. No prioritization may make all issues appear equally important. A missing recommendation may leave executives uncertain. Missing business impact may cause owners to reject or delay necessary spending.

The purpose of reporting is not to prove technical expertise. The purpose is to guide action. A report without a recommendation is incomplete because it transfers the burden of interpretation to people who may not understand the technical risk.

Good Engineering Report	Weak Engineering Report
Clear description in simple language.	Long technical descriptions with no context.
Risk level identified as low, medium, or high.	No prioritization between issues.

Cost of inaction explained.	No explanation of what happens if the hotel waits.
Recommended action included.	No recommendation or decision request.
Timing urgency stated.	No deadline or unclear urgency.
Business impact linked to guest, cost, safety, or asset value.	Technical readings without business interpretation.

Engineering KPIs Managers Should Understand

A KPI is only useful if it helps managers make better decisions. Some engineering KPIs speak directly to hotel performance. Others may be technically interesting but weak from an executive perspective. The goal is not to measure everything; it is to measure what supports operational reliability, guest protection, cost control, and risk reduction.

Useful engineering KPIs include the number of guest-impacting incidents, the ratio between preventive and corrective maintenance, energy cost per occupied room, equipment downtime hours, response time for urgent incidents, recurrence of the same technical issue, rooms out of order due to technical problems, and completion rate of preventive maintenance tasks.

Less useful KPIs include total number of work orders without classification, technical readings without context, equipment age without risk assessment, or maintenance hours logged without outcome. These figures may have operational value for the engineering team, but they do not automatically help management understand risk or performance.

The preventive versus corrective ratio is especially important. A hotel with a high level of corrective maintenance is likely operating reactively. A hotel with strong preventive completion and fewer emergency incidents is building operational stability. This KPI can help General Managers defend preventive budgets because it shows whether the hotel is moving toward control or crisis.

Energy cost per occupied room is also powerful because it connects engineering performance to financial performance. A hotel may have high energy consumption due to poor zoning, inefficient equipment, guestroom control problems, old systems, or weak monitoring. If tracked properly, the KPI helps management evaluate investments, training, and operational discipline.

Visual Reporting for Executive Committees

Executive reporting should be clear, concise, and decision-oriented. The objective is not to show how much data Engineering has collected. The objective is to show what leaders need to know.

A useful rule is one page, maximum clarity, minimum clutter. Traffic-light logic can help: green for controlled, yellow for watch, red for critical. Trends are usually more useful than isolated details because executives need to know whether performance is improving, declining, or stable. A single month of energy cost may be interesting; a six-month upward trend is a decision signal.

Executive reporting should always point toward action. If a dashboard shows a red indicator, it should also state what action is recommended. If a trend is deteriorating, it should explain what risk is emerging. If a preventive maintenance backlog is increasing, it should identify whether the cause is staffing, budget, parts availability, contractor delay, or poor scheduling.

The best engineering dashboards do not replace discussion. They structure discussion. They allow managers to focus on the right questions: What is critical? What is the trend? What is the cost of waiting? What decision do we need to take?

Executive Briefing: From Incident to Decision

The final skill in this chapter is preparing an executive briefing. A briefing is not a technical explanation. It is a short, structured message designed to help decision-makers understand a situation and act.

The mandatory structure used in this course includes five parts: context, risk, options, recommendation, and decision required. This format prevents vague communication. It also limits technical over-explanation and forces the speaker to take a managerial position.

Context explains what is happening. It should be short and free of jargon. Risk explains what may happen if the hotel waits or underreacts. Options should be limited, ideally to a maximum of two choices, because too many options create confusion. The recommendation states the preferred managerial direction. The decision required tells the audience exactly what is needed: budget approval, operational authorization, contractor engagement, temporary room closure, guest communication, or escalation to ownership.

A strong briefing sounds like this: "We have repeated temperature instability affecting 18 rooms on the east wing during evening occupancy. If we do not intervene, we risk continued guest complaints, compensation, and possible room blocks during the next high-demand period. We have two options: temporary adjustment and monitoring for three days, or immediate contractor inspection with potential overnight intervention. I recommend the contractor inspection because the complaints are recurring and the next occupancy peak is approaching. I need approval for the callout and authorization to block up to six rooms if required."

This briefing is not technical, but it is precise. It gives the executive team enough information to decide. It also demonstrates confidence, business logic, and ownership.

Common Communication Mistakes

Several mistakes weaken engineering communication. The first is excessive technical detail. Technical accuracy matters, but an executive briefing should not become an engineering lecture. The second mistake is no recommendation. Managers are not only messengers; they must help shape decisions. The third mistake is unclear urgency. If the audience does not know whether the issue must be solved today, this week, or this quarter, the briefing has failed.

Another mistake is missing financial context. Not every issue requires exact cost information immediately, but the speaker should identify whether the issue creates cost exposure, revenue exposure, or future capital risk. A final mistake is failing to explain guest impact. In hospitality, technical problems become meaningful when they affect comfort, safety, service continuity, or trust.

Building Credibility with Engineers, Owners, and Operations

Engineering vocabulary helps managers build credibility in three directions. With engineers, it shows respect for technical reality and allows more precise discussion. With owners, it allows the manager to defend investment in business terms. With operations teams, it helps explain why temporary disruption may be necessary to protect long-term service quality.

Credibility does not come from pretending to be an engineer. It comes from asking disciplined questions and translating the answer responsibly. A manager may say, "I do not need the technical drawing, but I need to understand the failure scenario, guest exposure, cost of inaction, and decision timeline." This statement is not technical, but it is managerial.

Over time, this communication approach improves hotel culture. Engineering feels heard because its risks are translated properly. Operations understands why planned maintenance matters. Finance sees the link between spending and risk reduction. Owners receive clearer business cases. The hotel moves from technical confusion to operational alignment.

Operational Vocabulary Beyond Maintenance

Maintenance vocabulary is the foundation, but managers also need a broader operational vocabulary to participate confidently in daily hotel decisions. This vocabulary does not require technical mastery. It requires understanding what each word means for operations, cost, guest experience, safety, and responsibility.

One important term is asset. In engineering discussions, the hotel is not only a place of service; it is a physical asset that must be preserved. Every guestroom, façade, roof, pump, elevator, boiler, corridor, kitchen hood, door, and technical room contributes to the value of the property. When maintenance is deferred, the hotel may protect short-term cash but weaken long-term asset value.

Another important term is lifecycle. Equipment has a life expectancy. It can be maintained, repaired, overhauled, or replaced, but it cannot be expected to perform indefinitely without investment. A manager who understands lifecycle thinking can distinguish between a small repair and a system approaching the end of useful life. This distinction matters because repeated small repairs can become more expensive than planned replacement.

Managers should also understand the idea of critical equipment. Not all equipment failures carry the same weight. A decorative lighting issue in a back corridor is not the same as a generator failure, a fire alarm fault, a main boiler instability, or a kitchen extraction problem. Critical equipment is equipment whose failure can threaten safety, business continuity, guest experience, or legal compliance.

Service level is another useful concept. It defines the expected level of performance and response. In a luxury hotel, for example, acceptable response times for guestroom temperature complaints may be very different from those in an economy property. Engineering strategy must therefore align with brand positioning. A five-star hotel cannot manage technical response with the same tolerance as a budget hotel if the guest promise is different.

Acronyms and the Risk of False Understanding

Engineering departments often use acronyms because they are efficient within technical teams. HVAC, BMS, AHU, VFD, PMS, UPS, and many other terms may appear in reports, meetings, or contractor documents. Acronyms are not a problem when everyone understands them. They become dangerous when managers pretend to understand them but do not connect them to risk.

A manager should never hesitate to ask for translation. Asking, "What does this mean for the guest, the operation, or the budget?" is more useful than pretending to understand a technical abbreviation. The goal is not to eliminate engineering language. The goal is to ensure that technical language does not block decisions.

For example, an AHU, or air handling unit, may sound like a technical box. In managerial language, it may represent fresh air supply, comfort stability, odor management, air quality, and guest perception. A UPS, or uninterrupted power supply, may sound like an IT or electrical detail. In business terms, it may protect front desk systems, payment continuity, server stability, access control, or security equipment during short interruptions.

The practical rule is simple: every acronym used in an executive meeting must be translated into impact. If it does not affect the decision, remove it. If it does affect the decision, explain why.

Technical Term or Acronym	**Plain Meaning for Managers**	**Typical Business Question**
HVAC	Heating, ventilation, and air conditioning systems that control comfort and air quality.	How does performance affect guest comfort, complaints, and energy cost?
BMS	Digital system used to monitor or control building systems.	Does the system help us reduce risk, improve consistency, or save energy?
AHU	Air handling unit supplying and conditioning air for zones of the building.	Which areas are affected if this unit fails?
UPS	Backup power support for short interruptions before generators or main power stabilize.	Which critical systems are protected, and for how long?
SLA	Service level agreement defining expected service or response standards.	What response time and performance level are we contractually expecting?

Prioritizing Engineering Issues

A common managerial challenge is prioritization. Engineering teams may present several issues at once: a leaking pipe, an overdue inspection, a noisy fan coil, a failing sensor, a guestroom temperature complaint, a kitchen equipment fault, and a proposal for energy optimization. Without prioritization logic, the Executive Committee may either delay everything or approve the loudest issue rather than the most important one.

Prioritization should combine impact and urgency. Impact asks what happens if the issue worsens. Urgency asks how quickly a decision is required. A high-impact, high-urgency issue requires immediate action. A high-impact but lower-urgency issue requires planning and budget protection. A low-impact issue may be scheduled without crisis language.

Managers should also distinguish between visible urgency and hidden criticality. A guest complaint may be visible and emotionally urgent, but a hidden safety compliance issue may be more critical. Strong managers do not only react to the most visible pain. They protect the hotel from the risks that may not yet be visible to guests.

A practical prioritization conversation can use three questions: Is anyone unsafe? Are guests currently affected? Will waiting make the issue more expensive, more disruptive, or less controllable? If the answer to any of these questions is yes, the issue must be escalated clearly.

Priority Category	Meaning	Managerial Response
Critical	Safety, compliance, major guest impact, or business continuity at risk.	Immediate escalation, clear authority, documented decision.
High	Significant operational or guest exposure, but manageable with prompt action.	Schedule urgent intervention and communicate to affected departments.
Medium	Limited impact or early warning indicator requiring planning.	Monitor, plan, budget, and review trend.
Low	Minor issue with limited operational exposure.	Schedule within routine maintenance planning.

Communicating with Owners and Asset Managers

Engineering communication changes when the audience changes. A Chief Engineer may need technical detail. A Rooms Division Manager may need guest impact. A Financial Controller may need cost exposure. An owner or asset manager needs to understand the relationship between investment, risk, and asset value.

When speaking to owners, managers should avoid presenting engineering as a list of repairs. Owners are more likely to respond to structured business cases. A business case should explain why the issue matters, what risk the owner currently carries, what options exist, what investment is recommended, and how the decision protects value.

For example, a request to replace guestroom fan coil units may be rejected if presented as a technical upgrade. It becomes stronger when presented as a pattern of recurring comfort complaints, rising corrective maintenance cost, energy inefficiency, room downtime, and brand promise risk. The same investment can then be understood as guest experience protection and asset improvement.

Asset managers also expect managers to distinguish between OPEX and CAPEX logic. Some interventions are routine operating expenses. Others are capital investments that extend life, improve value, or replace major systems. Even if this distinction is developed further in the next chapter, the communication skill begins here: managers must be able to explain whether a cost is maintaining current performance or improving the asset for the future.

Communicating with Contractors and Specialists

Hotels often rely on external contractors for specialized systems such as elevators, fire safety, kitchen extraction, generators, chillers, building management systems, and water treatment. Communication with contractors requires discipline because outsourcing execution does not outsource responsibility.

Managers should ensure that contractor communication includes scope, timing, access requirements, guest impact, safety measures, documentation, and follow-up. A contractor visit is not complete simply because the technician leaves the site. The hotel needs a clear report, findings, recommendations, and any compliance certificates or service records required for future reference.

A weak contractor relationship can create dependency without clarity. A strong relationship creates expert support while preserving hotel accountability. Managers should ask contractors to present recommendations in a way that supports decisions: what is mandatory, what is recommended, what can wait, what risk exists if the hotel delays, and what documentation will be provided.

The Communication Chain During an Incident

During an engineering incident, communication must move quickly but remain controlled. Too many voices create confusion. Too little information creates panic. The manager must establish a clear communication chain.

The first communication should confirm what is known and what is not yet known. Speculation should be avoided. The second communication should identify guest impact and operational impact. The third should clarify immediate actions and decision ownership. The fourth should provide updates at agreed intervals. The final communication should document root cause, actions taken, cost, guest compensation if any, and preventive follow-up.

This sequence protects the hotel in several ways. It supports operational coordination, reduces inconsistent messages, helps Front Office respond to guests, supports Finance and ownership reporting, and creates documentation for future review. In serious cases, it may also support insurance, legal, or regulatory processes.

Communication after the incident is equally important. A post-incident review should ask what happened, why it happened, how it was detected, how quickly the hotel responded, what communication worked, what failed, what costs were incurred, and what must change to prevent recurrence. Without this review, the hotel may repair the symptom but keep the weakness.

Creating a Culture of Engineering Clarity

Engineering communication is not only an individual skill. It is also a culture. Hotels with a strong engineering culture do not wait for crisis to discuss technical risk. They include engineering in operations meetings, review preventive maintenance performance, connect guest complaints to technical patterns, and make asset condition visible to senior leaders.

This culture requires mutual respect. Operations should not treat Engineering only as a repair service. Engineering should not hide behind jargon. Finance should not see maintenance only as cost. The General Manager should create a shared language where each department understands how technical systems support the guest promise.

A practical way to build this culture is to require every significant engineering item to include the same five elements: context, risk, impact, recommendation, and decision required. Over time, this structure trains the organization to move away from vague problem reporting and toward disciplined decision-making.

When a hotel builds this culture, engineering becomes less reactive and more strategic. The hotel sees patterns earlier. Budgets become easier to defend. Owners receive better explanations. Guests experience fewer disruptions. Staff feel less exposed to crisis. The asset is protected more intelligently.

CHAPTER SUMMARY

Engineering vocabulary is a managerial tool. It allows future hotel leaders to understand engineering issues without becoming technical specialists. The chapter has shown that engineering language must be translated into business impact, because owners and executives fund risk reduction, guest experience protection, cost control, compliance, and asset preservation.

The key maintenance categories are corrective, preventive, predictive, planned, and reactive interventions. Corrective maintenance responds after failure. Preventive maintenance reduces the chance of failure. Predictive maintenance uses data to anticipate risk. Planned interventions create control. Reactive interventions expose the hotel to pressure, cost, and visible disruption.

The manager's role is not to repair equipment. The manager's role is to ask the right questions, prioritize, defend budgets, and explain technical issues to decision-makers. This requires a clear translation framework: what is the issue, what happens if we do nothing, who is impacted, and what decision is required.

Engineering reports and KPIs must support decisions. Strong reports include risk level, cost of inaction, recommended action, and urgency. Strong KPIs focus on guest-impacting incidents, preventive versus corrective ratio, energy cost per occupied room, equipment downtime, and recurrence of issues. Weak KPIs provide data without context.

Finally, executive briefings must be concise and decision-oriented. The recommended structure is context, risk, options, recommendation, and decision required. A good hospitality manager does not hide behind technical jargon. A good manager translates engineering reality into clear operational, financial, and strategic choices.

OPERATIONAL CASE STUDY

The Silent Boiler Warning

The Bellecour Regency is a 185-room four-star hotel located in a busy European city. The property includes a breakfast restaurant, a small spa, three meeting rooms, and a high percentage of corporate guests during the week. The hotel has recently improved its online reputation after several months of operational instability. Occupancy is expected to reach 92% next week due to a citywide trade fair.

During the Monday morning operations meeting, the Chief Engineer informs the Executive Committee that the main hot water boiler has shown pressure instability during peak morning demand. The engineering team has adjusted the settings twice over the past ten days. No full outage has occurred yet, but the system is no longer performing consistently. The Chief Engineer recommends bringing in a specialist contractor for inspection and possible parts replacement before the trade fair begins.

The Financial Controller asks whether the intervention can wait until next month because the hotel is already under pressure to reduce operating expenses. The Rooms Division Manager argues that there are no major complaints yet, so the issue should not disturb next week's occupancy. The General Manager asks for more information, but the first explanation is technical and unclear: "The boiler pressure is fluctuating due to a probable valve or control issue linked to demand peaks."

The meeting becomes confused. The issue is not properly translated. Some managers hear "possible technical problem." Others hear "budget request." Nobody has yet framed the situation as a guest experience, revenue, and reputation risk.

Managerial Analysis

The first issue is not the boiler. The first issue is communication. The technical warning has not been expressed in a decision-ready format. The Executive Committee needs to know what could happen if the hotel waits, how many guests may be impacted, what the timing risk is, what options exist, and what decision must be taken.

Using the translation framework, the issue can be reframed as follows. The hotel is facing recurring instability in the hot water production system during peak demand periods. If no action is taken before the trade fair, the hotel may experience inconsistent hot water during morning hours when occupancy is high. The guests most affected would be corporate guests preparing for meetings and event attendees with limited time. The business impact could include complaints, compensation, poor reviews, loss of trust, staff pressure, and possible room revenue protection measures. The decision required is whether to approve an immediate specialist inspection and authorize preventive intervention before the occupancy peak.

The issue is therefore not simply a technical request. It is a risk arbitration decision. The hotel can delay and accept the risk, or it can intervene now and protect the guest experience. A strong manager would ask the Chief Engineer to quantify the exposure: how many rooms could be impacted, when the risk is highest, whether temporary measures exist, whether parts are available, and what the estimated cost range is. The manager would also ask Finance to compare intervention cost with possible compensation, reputational damage, and operational disruption.

Recommended Executive Briefing

<u>Context</u>: "We have a recurring hot water stability issue affecting the main boiler during peak morning demand. No full outage has

occurred yet, but the pattern has appeared twice in ten days and next week's occupancy is forecast above 90%."

Risk: "If we wait, we risk hot water inconsistency during the trade fair, affecting corporate guests at the most sensitive moment of the day. This could create complaints, compensation, negative reviews, and operational pressure at Front Office."

Options: "Option one is to monitor and defer inspection until next month, accepting the risk during high occupancy. Option two is to approve a specialist inspection this week and authorize preventive parts replacement if required."

Recommendation: "I recommend option two because the risk period is immediate, the guest impact would be highly visible, and preventive intervention is likely less costly than a crisis during peak occupancy."

Decision required: "Approval is requested today for the contractor inspection and a provisional budget envelope for urgent corrective-preventive work before Friday."

Lessons from the Case

This case illustrates the difference between technical reporting and executive communication. The original statement was technically useful but managerially incomplete. The reframed briefing creates clarity, identifies risk, connects the issue to guest experience, presents options, recommends action, and asks for a decision.

The case also shows why preventive maintenance can be difficult to defend. When the system has not yet failed, some managers may prefer to delay. But in hospitality, waiting until failure often means allowing the guest to discover the weakness. Strong managers understand that the best engineering decisions are often invisible because they prevent incidents before guests experience them.

CHAPTER KEY TERMS

- **Corrective maintenance:** Repair or intervention performed after equipment or a system has already failed. It is often urgent, visible to guests, and more expensive than planned work.
- **Preventive maintenance:** Planned maintenance designed to reduce the probability of future failure. It protects guest experience, asset value, safety, and cost control.
- **Predictive maintenance:** Maintenance approach using data, monitoring, trends, or sensors to anticipate problems before visible failure occurs.
- **Planned intervention:** Maintenance or repair scheduled in advance to control timing, cost, guest impact, staffing, and communication.
- **Reactive intervention:** Emergency response after an unexpected failure or complaint. It often indicates that earlier preventive action may have been insufficient.
- **Downtime:** Period during which equipment, a room, an outlet, or a system cannot operate normally. Downtime can create guest dissatisfaction and revenue loss.
- **Criticality:** Degree to which a system or piece of equipment is essential to safety, guest experience, operations, revenue, or compliance.
- **Cost of inaction:** Financial, operational, reputational, or legal consequence of delaying a decision or intervention.

- **Guest-impacting incident:** Technical issue that directly affects guest comfort, safety, access, service quality, or perception of the hotel.
- **Preventive versus corrective ratio:** KPI comparing planned preventive work with reactive corrective work. It indicates whether the hotel is operating proactively or reactively.
- **Energy cost per occupied room:** KPI connecting energy consumption to occupancy. It helps managers evaluate efficiency and cost performance.
- **Equipment downtime hours:** Total time during which essential equipment is not operational. This KPI helps measure operational reliability.
- **BMS:** Building Management System. A system used to monitor and control technical installations such as HVAC, energy use, alarms, or equipment performance.
- **Risk level:** Assessment of seriousness, often expressed as low, medium, or high, based on probability and impact.
- **Executive briefing:** Short, structured communication designed to help decision-makers understand an issue and approve an action.
- **Decision required:** Clear statement of what approval, authorization, budget, or action is needed from management.

Chapter Review Questions

- Why is engineering vocabulary a source of managerial authority rather than only technical language?
- Explain the difference between corrective maintenance and preventive maintenance using a hotel example.

- Why can preventive maintenance look expensive even when it protects the hotel financially?
- What does it mean to translate a technical problem into business impact?
- Using the translation framework, how would you present an overdue kitchen hood inspection to an Executive Committee?
- Why is a report without a recommendation incomplete from a managerial perspective?
- Which engineering KPIs are most useful for a hotel General Manager, and why?
- How can a high corrective maintenance ratio indicate deeper organizational weakness?
- Why should executive engineering reporting focus on trends and decisions rather than technical detail?
- Prepare a short executive briefing structure for an elevator breakdown affecting guest access during peak check-in.

CHAPTER REFERENCES

- ASHRAE. (2023). ASHRAE Handbook: HVAC Applications. American Society of Heating, Refrigerating and Air-Conditioning Engineers.
- Atkin, B., & Brooks, A. (2021). Total Facility Management. Wiley-Blackwell.
- International Facility Management Association. (2022). Operations and Maintenance Benchmarks. IFMA.
- Kandampully, J., Zhang, T., & Jaakkola, E. (2018). Customer experience management in hospitality: A literature synthesis,

new understanding and research agenda. International Journal of Contemporary Hospitality Management.

- Ransley, J., & Ingram, H. (2012). Developing Hospitality Properties and Facilities. Routledge.
- Stipanuk, D. M. (2015). Hospitality Facilities Management and Design. Educational Institute of the American Hotel & Lodging Association.
- Walker, J. R. (2021). Introduction to Hospitality Management. Pearson.
- Course materials: C4 - Engineering Vocabulary and Operational Communication, presentation and teacher notes.
- International Organization for Standardization. (2018). ISO 41001: Facility management - Management systems - Requirements with guidance for use.
- International Organization for Standardization. (2014). ISO 55000: Asset management - Overview, principles and terminology.
- International Organization for Standardization. (2018). ISO 50001: Energy management systems - Requirements with guidance for use.
- National Fire Protection Association. (2024). NFPA 70: National Electrical Code. NFPA.
- CIBSE. (2014). Guide M: Maintenance Engineering and Management. Chartered Institution of Building Services Engineers.
- RICS. (2021). Strategic Facilities Management: Guidance Note. Royal Institution of Chartered Surveyors.
- BOMA International. (2020). Preventive Maintenance Guidebook: Best Practices to Maintain Efficient and Sustainable Buildings.

Chapter 5
Managing Renovation Programs: CAPEX & OPEX

Engineering Projects & Financial Logic for Hospitality Managers

CHAPTER GENERAL OVERVIEW

Hotel renovations are often discussed as design, construction, or engineering projects. For a future hotel manager, this view is incomplete. A renovation decision is also a financial decision, an operational risk decision, a guest experience decision, and a leadership decision. The central managerial question is not only whether a system can be repaired, replaced, modernized, or postponed. The real question is what risk the hotel accepts when it delays investment, how much operational disruption it can tolerate, and how convincingly the executive team can defend its choices to owners, finance, employees, guests, and sometimes regulators.

This chapter develops the managerial logic required to understand engineering renovation programs through the lens of CAPEX and OPEX. Students do not need to become engineers, architects, quantity surveyors, or finance specialists. They do, however, need to understand how technical deterioration becomes financial exposure; how maintenance decisions can mask deferred investment; how renovation timing can either protect or destroy value; and how visible improvements must be weighed against invisible but critical infrastructure needs.

The chapter follows the same executive logic used throughout the course. Chapter 1 established that engineering is the foundation of business continuity. Chapter 2 explained that infrastructure is directly connected to guest experience. Chapter 3 showed that safety is a leadership responsibility. Chapter 4 developed the communication skills needed to translate technical issues into business impact. Chapter 5 now brings these elements together through renovation planning and financial arbitration.

The guiding sentence for this chapter is simple: renovation projects often fail financially before they fail technically. This does not mean that engineering expertise is secondary. It means that even technically competent projects can become damaging when they are poorly phased, under-budgeted, weakly communicated, or disconnected from the hotel's operating cycle. A chiller replacement, a bathroom renovation, an electrical compliance upgrade, or a full HVAC modernization may all make technical sense. The managerial challenge is to determine what must be done first, what can be delayed, what delay will cost, and how the decision should be explained.

CHAPTER CONTENT

Renovation as an Executive Decision

A renovation program begins long before contractors arrive on site. It begins when the hotel recognizes that a system, space, or asset no longer supports the required level of performance. The trigger may be technical, such as repeated equipment breakdowns or end-of-life infrastructure. It may be commercial, such as declining guest satisfaction, reduced rate competitiveness, or a repositioning strategy. It may be regulatory, such as fire safety, accessibility, energy performance, or building code requirements. It may also be brand-driven, especially when the property belongs to a chain or is required to comply with updated standards.

From the managerial perspective, renovation is never neutral. It consumes capital, affects cash flow, disrupts operations, requires coordination across departments, and changes the risk profile of the property. A poorly timed renovation can reduce occupancy during a period when demand is strong. A poorly communicated renovation can generate guest complaints even when the work itself is necessary. A poorly scoped renovation can solve visible problems while leaving the underlying infrastructure exposed. A poorly defended renovation budget can lead ownership to delay investment until the cost of inaction becomes higher than the cost of prevention.

This is why the executive committee must approach renovation not as a technical wish list but as a structured decision-making process. The question is not "What does Engineering want?" but "What does the asset require, what risks are we carrying, what return or protection does the investment provide, and what happens if we do nothing?"

Why Hotels Renovate

Hotels renovate for several reasons, and the reason behind a renovation strongly affects how it should be prioritized and financed. The first driver is asset aging. Technical systems such as HVAC, boilers, chillers, pumps, electrical panels, plumbing networks, elevators, and fire safety components have useful lives. They can be maintained, optimized, and repaired, but they cannot last indefinitely. When equipment reaches the end of its lifecycle, maintenance alone may become financially irrational.

The second driver is brand standards. A property may need to renovate because the brand has evolved, the competitive set has improved, or the hotel is being repositioned. In this case, renovation may include both visible elements, such as bathrooms, rooms, lobbies, restaurants, and meeting spaces, and invisible elements, such as technical upgrades required to support the new level of service.

The third driver is regulatory compliance. Some renovations are not optional. Fire and life safety systems, accessibility obligations, building permits, energy performance requirements, electrical compliance, and hygiene-related infrastructure can force investment even when ownership would prefer to delay. A manager must recognize when a project is no longer a choice but a business survival requirement.

The fourth driver is market pressure. Guest expectations evolve. Competitors invest. Online reviews make technical shortcomings visible. A hotel that postpones renovation too long may experience rate resistance, declining guest satisfaction, and lower market relevance. In this context, renovation becomes a way to protect revenue, not simply a way to improve the physical product.

Renovation Driver	**Typical Signal**	**Managerial Meaning**
Asset aging	End-of-life systems, repeated failures, declining reliability	Risk of escalating OPEX and operational disruption
Brand standards	Repositioning, franchise requirements, updated guest promise	Need to align physical product with market positioning

Regulatory compliance	Safety, accessibility, energy, permits, building codes	Mandatory investment; delay can create legal or insurance exposure
Market pressure	Competitor upgrades, guest expectations, review decline	Revenue protection and brand competitiveness

Renovation Is a Lifecycle, Not a Single Project

One of the most important mistakes in hotel renovation is to treat the project as a construction event rather than as a lifecycle. A renovation lifecycle begins with diagnostic work and ends with post-renovation evaluation. The hotel that skips these stages often pays for it through cost overruns, poor sequencing, operational disruption, and incomplete learning.

The first stage is diagnostic and audit. This includes technical audits, asset condition surveys, energy performance reviews, compliance checks, guest feedback analysis, and operational pain-point mapping. The purpose is not only to identify what is broken, but to understand why it is happening, what the consequences are, and how urgent the intervention is.

The second stage is strategic decision-making. At this stage, the executive committee clarifies why the project is needed, what should be done, when it should happen, and what level of risk is acceptable. A renovation decision should never be reduced to a single price. It must also include timing, phasing, guest impact, compliance exposure, cash availability, and long-term asset value.

The third stage is budgeting and phasing. This is where many projects fail. A budget that only includes contractor costs is incomplete. The hotel must include room revenue loss, temporary closures, guest compensation, communication needs, staff overtime, project management time, testing, commissioning, contingency, and sometimes energy inefficiency during partial operations. Phasing determines whether the property can continue operating without damaging guest trust.

The fourth stage is execution. During execution, coordination becomes critical. Rooms Division must manage inventory. Front Office must manage expectations. Sales and Revenue must manage pricing and communication. Housekeeping must adapt to room closures and contractor access. Security and Engineering must control safety. Finance must monitor cash flow. The General Manager must maintain alignment between the owner's expectations and operational reality.

The fifth stage is commissioning and post-impact evaluation. Commissioning means verifying that systems work as intended before they are fully handed over to operations. Post-renovation evaluation means checking whether the investment delivered its expected results: lower energy costs, fewer guest complaints, improved reliability, better compliance, higher rates, or stronger asset value. Many hotels underestimate this stage and lose the opportunity to measure whether the investment actually achieved its purpose.

Engineering Renovation versus Cosmetic Renovation

Not all renovation projects carry the same type of value or risk. A cosmetic renovation improves what the guest sees: room design, furniture, finishes, bathroom aesthetics, lobby atmosphere, restaurant ambiance, or public area identity. Cosmetic renovation can support marketing, brand repositioning, guest perception, and rate strategy. It is often visible, emotionally appealing, and easier for non-technical stakeholders to understand.

An engineering renovation improves what the guest usually does not see: HVAC systems, plumbing networks, boilers, chillers, water pressure, electrical panels, emergency power, fire safety systems, ventilation, insulation, control systems, or building distribution networks. These investments may not photograph well, but they protect the business. They reduce failure risk, protect compliance, improve energy performance, safeguard asset integrity, and support service continuity.

The danger in renovation arbitration is that visible projects often win emotional support while invisible projects carry the greater risk. Owners, commercial teams, and even operators may prefer bathroom upgrades because they appear more directly connected to guest satisfaction. However, if the electrical system is non-compliant or the HVAC system is at the end of life, the hotel may be prioritizing appearance over survival. The correct managerial approach is not to reject cosmetic investment, but to rank projects according to risk, compliance, guest impact, financial logic, and asset protection.

Type	**Description**	**Strategic Value**
Cosmetic renovation	Visible improvements to rooms, bathrooms, lobby, restaurant, or design identity	Short-term guest perception, marketing appeal, rate support
Engineering renovation	Invisible infrastructure such as HVAC, plumbing, electrical, boilers, chillers, safety systems	Long-term reliability, compliance, energy efficiency, asset protection

Integrated renovation	Combines visible upgrades with infrastructure modernization	Often the strongest approach, but requires stronger planning and higher coordination

Stakeholders and Decision-Makers

Renovation decisions rarely belong to one department. Engineering understands technical feasibility and system requirements. Operations understands guest disruption and service continuity. Finance understands cash flow, depreciation, and P&L implications. Ownership or asset management focuses on return on investment, asset valuation, and long-term strategy. The difficulty is that these stakeholders often speak different languages.

Engineering may say: "The chiller is inefficient and the system is close to failure." Finance may ask: "What is the cash impact this year?" Operations may ask: "How many rooms will be out of order during high season?" Ownership may ask: "Will this increase value or only prevent loss?" The manager's role is to connect these perspectives into a decision that is technically sound, operationally realistic, financially coherent, and strategically defensible.

The executive committee must therefore build a shared language. Engineering needs to translate technical requirements into risk, cost, timing, and business impact. Finance needs to understand that deferring CAPEX can create rising OPEX and hidden losses. Operations needs to understand why a temporary disruption may be necessary to prevent a larger failure. Ownership needs to understand that some projects protect value rather than immediately create new revenue.

CAPEX and OPEX Logic

CAPEX, or capital expenditure, refers to long-term investment in the asset. It is usually capitalized and depreciated over time. In engineering renovation, CAPEX may include replacing the HVAC system, upgrading electrical infrastructure, replacing boilers and chillers, renewing major plumbing networks, modernizing fire safety infrastructure, or performing structural upgrades. CAPEX normally requires ownership or board approval because it has a significant upfront cash impact and affects the asset's long-term value.

OPEX, or operating expenditure, refers to recurring operating costs that affect the profit and loss statement more immediately. In engineering, OPEX includes maintenance contracts, temporary repairs, emergency call-outs, recurring outsourcing, spare parts, energy overconsumption, and ongoing operational fixes. OPEX may be easier to approve in the short term, but it can become financially dangerous when it is used to postpone necessary capital investment.

The managerial danger is the illusion that postponing CAPEX saves money. In some cases, it does. A project may be delayed because the asset still performs adequately, the risk is low, the market cycle is unfavorable, or cash must be protected. But postponing CAPEX without understanding the consequences can create an OPEX trap. Temporary repairs multiply. Energy waste continues. Staff spend time firefighting. Guest complaints increase. Emergency interventions cost more than planned work. The asset loses reliability. In these cases, OPEX is not a true saving; it is deferred CAPEX in disguise, often at a higher total cost.

Concept	Meaning	Hotel Example
CAPEX	Long-term asset investment; capitalized and depreciated over time	HVAC replacement, electrical upgrade, boiler replacement, full plumbing renewal
OPEX	Operating expense affecting current operating performance	Maintenance contracts, emergency repairs, energy overuse, temporary patches
Deferred CAPEX	Capital investment postponed beyond the rational point	Old system repeatedly repaired instead of replaced
OPEX trap	Operating cost pattern caused by delayed capital investment	Frequent breakdowns, high call-out costs, guest compensation, energy waste

Deferred Maintenance and the Cost of Waiting

Deferred maintenance occurs when necessary repairs, replacements, or upgrades are postponed. It can be intentional, because the hotel lacks capital or chooses to prioritize another project. It can also be unintentional, because the property has no reliable asset register, no preventive maintenance discipline, or no executive understanding of infrastructure risk. In both cases, deferred maintenance accumulates exposure.

The cost of waiting is rarely limited to the repair itself. A failing HVAC system may produce guest complaints, refunds, negative reviews, poor sleep quality, and energy overconsumption. A failing plumbing network may lead to leaks, room closures, water damage, hygiene concerns, and compensation. Non-compliant electrical panels may create safety exposure and insurance concerns. An old boiler may require emergency intervention at exactly the wrong moment. The financial cost is therefore not only the technical repair. It includes lost room revenue, guest dissatisfaction, staff stress, owner frustration, and reputational damage.

For managers, the key discipline is to compare the total cost of action with the total cost of inaction. The price of a renovation quote may appear high, but the cost of doing nothing may be higher. A credible executive briefing should therefore include not only project cost, but also risk exposure, expected operational impact, hidden OPEX, potential revenue loss, compliance consequences, and timing options.

Budget Planning and Phasing

Budget planning is not simply the act of collecting quotes. It is the act of translating strategic priorities into a financially and operationally executable plan. A strong renovation budget includes technical cost, professional fees, permits, contingency, procurement, temporary equipment, testing, commissioning, communication, internal project management time, and operational disruption. A weak budget includes only the visible contractor price and then discovers the rest too late.

Phasing is equally important. Phasing is the sequential implementation of renovation work in order to protect cash flow, guest experience, staff workload, safety, and operational continuity. A hotel may phase work by floor, wing, department, system, season, or business priority. The correct phasing strategy depends on demand patterns, occupancy forecasts, event calendars, contractor availability, cash reserves, brand requirements, and the level of risk accepted by ownership.

Good phasing protects financial survival. It aligns work with low-occupancy periods, minimizes room displacement, allows staff to adapt, and gives the commercial team time to communicate honestly with guests and clients. Poor phasing can destroy value even when the renovation itself is justified. A project scheduled during peak demand may generate unnecessary revenue loss. A project executed without clear communication may create review damage. A project split into too many phases may increase total cost and fatigue the team. A project compressed too aggressively may create quality defects or safety risks.

Hidden Costs Nobody Budgets

Renovation budgets are often underestimated because hidden costs sit between CAPEX and OPEX. They may not appear in the original construction quote, but they affect the hotel's financial performance. The first hidden cost is room revenue loss. If rooms are closed for works, they cannot be sold. If floors are partially closed, housekeeping and maintenance efficiency may decline. If public areas are disrupted, groups or events may cancel or negotiate lower rates.

The second hidden cost is guest compensation. Noise, dust, delays, smell, elevator disruption, water interruptions, or temporary route changes may require service recovery. Even when the hotel communicates in advance, guests may demand discounts, loyalty points, upgrades, transfers, or refunds. The third hidden cost is energy inefficiency during construction phases. Temporary systems, partial operation, inefficient old equipment, or poor zoning can increase consumption.

The fourth hidden cost is staff overtime and stress. Renovation often requires employees to manage complaints, coordinate access, move supplies, inspect rooms, adjust procedures, and solve problems that are not part of normal operations. The fifth hidden cost is internal project management time. Senior leaders, department heads, Engineering, Housekeeping, Front Office, Sales, Finance, and Security all spend time managing the project. This time has a cost because it diverts attention from daily performance.

When managers ignore these hidden costs, renovation decisions appear cleaner than they really are. A credible business case must estimate them, explain them, and include contingency. This does not mean every number can be perfectly predicted. It means the executive committee recognizes that the real cost of renovation is broader than the contractor invoice.

Renovation During Operations

Hotels are difficult renovation environments because they remain open while work takes place. Unlike a closed construction site, a hotel must protect the experience of paying guests, the safety of employees, the continuity of events, and the reputation of the brand. This creates operational disruption risks that must be managed deliberately.

Guest complaints are the most immediate risk. Guests may tolerate some inconvenience if expectations are set clearly, but they rarely forgive surprises. Noise at the wrong time, blocked access, malfunctioning elevators, intermittent hot water, or visible construction disorder can turn a necessary renovation into a perception of poor management. Online reviews often judge the hotel not only for the renovation itself, but for how honestly and professionally it communicated the situation.

Revenue loss is another risk. Rooms may be taken out of order. Meeting spaces may become less attractive. Groups may require reassurance. Revenue management may need to adjust pricing strategy. Sales must decide whether to disclose ongoing works in proposals and contracts. Front Office must be prepared to answer questions consistently. Rooms Division must control inventory and ensure that renovated and non-renovated rooms are assigned appropriately.

Safety exposure must also be considered. Contractor movement, temporary barriers, dust, electrical work, hot works, water shutdowns, scaffoldings, delivery routes, and emergency egress must be controlled. A renovation that creates a safety incident damages the very purpose of improving the asset. Engineering and Security must therefore coordinate access, permits, supervision, and emergency procedures throughout the works.

Coordination with Finance

Finance is not simply the department that says yes or no to a budget. Finance helps the hotel understand timing, cash impact, depreciation, EBITDA implications, return expectations, and asset valuation. Engineering and Operations must therefore avoid presenting renovation needs in technical language only. A statement such as “the boiler is old” is not enough. Finance needs to understand what the old boiler costs, what risk it carries, what failure would mean, how replacement would affect cash, and what the depreciation timeline would be.

Typical finance questions include: What is the cash impact this year? What is the depreciation timeline? How does the investment affect EBITDA? What OPEX will be reduced if the CAPEX is approved? What revenue is protected? What is the risk of regulatory non-compliance? What is the impact on asset valuation? How does the project fit into the owner’s hold strategy or exit strategy?

A strong manager answers these questions in business language. Instead of saying "the electrical panels need upgrading," the manager says: "The electrical panels present a compliance and operational continuity risk. The investment protects the hotel from regulatory exposure, reduces emergency failure risk, and supports the next phase of room modernization. If postponed, we accept increased insurance scrutiny, potential disruption, and higher emergency costs." This is not jargon; it is executive translation.

When Renovation Becomes Mandatory

Some renovation decisions are discretionary. Others are not. A discretionary project may improve competitiveness, guest perception, design quality, or revenue potential. A mandatory project is required because the hotel is exposed to legal, safety, compliance, or catastrophic operational risk. Managers must be able to distinguish between these categories because the budget conversation changes completely.

Fire and life safety is a non-negotiable trigger. If sprinkler systems, fire alarms, emergency lighting, evacuation routes, or compartmentation are non-compliant, the project cannot be evaluated only through return on investment. Legal compliance is another trigger. Accessibility requirements, permits, building codes, and local regulations can force renovation. Energy regulations may also make certain upgrades unavoidable, especially as hotels face increasing reporting and sustainability obligations. System failure risk is another trigger. When equipment is close to catastrophic breakdown, delay may no longer be financially rational.

The executive rule is clear: mandatory projects are not optional investments; they are business survival requirements. The manager may still need to phase the work, negotiate timing, control cost, and communicate carefully, but the strategic choice is not whether safety or compliance matters. It does.

How to Build Renovation Arbitration Logic

Budget arbitration is the process of deciding between competing needs when resources are limited. In hotels, this is common. A property may need to replace HVAC, renovate bathrooms, upgrade electrical panels, repair back-of-house areas, modernize fire systems, and refresh the lobby, while only having enough capital for one or two projects. The managerial skill is to prioritize without being seduced only by visibility or emotion.

A strong arbitration process begins with compliance and safety. Any project linked to legal obligation, life safety, or serious operational continuity risk must be considered first. The second criterion is guest impact. If a technical issue is causing repeated complaints, compensation, poor reviews, or declining satisfaction, its commercial consequences must be quantified. The third criterion is financial impact. This includes CAPEX required, OPEX consequences if delayed, hidden costs, energy waste, and revenue protection. The fourth criterion is asset value. Some investments protect the long-term value of the property even when they do not immediately increase revenue. The fifth criterion is execution feasibility. A project may be important but impossible to execute during peak season or with current cash constraints.

The final decision should always state what risk is being accepted. Every budget choice postpones something. If the hotel funds electrical compliance first, it may accept another year of guest bathroom dissatisfaction. If it funds bathrooms first, it may accept technical risk that could prove unacceptable. If it funds HVAC first, it may protect comfort and energy performance but postpone visible repositioning. Mature executive decisions do not pretend there is no trade-off. They name the trade-off and defend it.

Arbitration Criterion	Managerial Question	Typical Priority
Compliance and safety	Is this required by law, insurance, code, or life safety obligation?	Very high
Guest impact	Does this issue affect comfort, satisfaction, reviews, rate resistance, or loyalty?	High
Financial impact	What are CAPEX needs, hidden OPEX, energy waste, compensation, and revenue loss?	High
Asset value	Does the project protect long-term property value and future sale or financing potential?	Medium to high
Execution feasibility	Can it be completed without unacceptable disruption, cash pressure, or quality risk?	Medium

Communicating Renovation Decisions to Owners

Owners and asset managers expect clarity. They do not want a technical complaint list. They want to understand risk, value, cash, timing, and decision requirements. A renovation recommendation should therefore be structured around a business case. It should explain the current issue, the risk of doing nothing, the options available, the recommended option, the financial logic, the operational impact, and the decision required.

The language must be precise but not technical. For example, instead of saying "The HVAC is 18 years old and inefficient," a manager might say: "The HVAC system is beyond its optimal lifecycle. We are seeing rising guest complaints and energy costs. Continuing to patch the system for two more years may appear cheaper, but it increases breakdown risk, guest compensation, and energy waste. A phased CAPEX replacement would protect comfort, reduce operating exposure, and allow us to control timing before failure forces an emergency intervention."

Strong communication also avoids false certainty. Renovation always contains uncertainty. Costs may change, contractor availability may shift, and operational disruptions may occur. A credible manager does not promise perfection. Instead, the manager demonstrates that risks have been identified, contingencies included, stakeholders aligned, and monitoring mechanisms established.

The final governance principle is escalation. Not every issue belongs at owner level, but some do. Delays affecting reopening, cost overruns beyond contingency, compliance risks, safety concerns, major guest disruption, and scope changes that affect asset value must be escalated quickly. A mature manager does not hide bad news. The manager frames it, explains options, recommends a decision, and protects the credibility of the hotel leadership team.

Communication governance is equally important. Guests do not need technical explanations, but they need honesty and consistency. Sales teams must know what can and cannot be promised to groups. Front Office must know how to respond to complaints. Housekeeping must know which rooms are affected and when. Revenue Management must understand inventory restrictions. Security must know contractor movements and emergency access constraints. Engineering must communicate shutdowns before they become guest incidents. The General Manager must ensure that all departments speak with one voice.

Change orders require particular attention. A change order is a modification to the original scope, cost, or timing of the project. Some change orders are necessary because hidden conditions are discovered once work begins. Others result from weak planning, unclear scope, or late stakeholder requests. Managers must not approve change orders casually. Each change should be evaluated according to business necessity, cost, delay, guest impact, and strategic value.

A basic governance rhythm should include a weekly project meeting, a risk register, a decision log, a cost tracker, an operational disruption plan, and a communication plan. The weekly meeting should not become a technical discussion only. It should review progress, upcoming disruptions, rooms or areas affected, safety issues, guest communication, budget movement, and decisions required. The risk register should identify what could go wrong and who owns the response. The decision log should record who approved changes, why they were approved, and what financial or operational consequence they carry.

Renovation governance is the system of meetings, reports, decisions, approvals, and escalation channels that keeps a project under control. In a hotel environment, governance must be strong because renovation affects guests, staff, revenue, safety, and ownership confidence at the same time. Without governance, projects drift. Decisions become informal. Departments receive information too late. Costs increase silently. Guest complaints become predictable but unmanaged.

Governance Discipline During Renovation

The General Manager should also ensure that new systems are transferred properly to the local engineering team. A new installation can fail to deliver value when staff are not trained, manuals are not stored, warranties are not tracked, or preventive maintenance schedules are not updated. Commissioning and handover are therefore managerial responsibilities. They ensure that investment becomes operational capability rather than a finished construction file.

Post-impact evaluation also protects the organization from repeating mistakes. The executive committee should ask whether the original budget was realistic, whether the phasing plan protected operations, whether communication to guests was adequate, whether contractors respected service constraints, whether internal teams were properly involved, and whether the contingency was sufficient. This learning is not administrative. It strengthens future renovation decisions and improves the hotel's ability to defend capital requests.

A strong post-renovation review compares the initial business case with the real outcome. If the project was justified by energy reduction, the hotel should track energy cost per occupied room before and after implementation. If it was justified by guest comfort, the hotel should monitor complaints, review language, room moves, compensation, and satisfaction scores. If it was justified by compliance, the hotel should keep inspection certificates, insurance correspondence, commissioning reports, and maintenance schedules in one accessible file. If it was justified by asset value, the hotel should record how the investment supports future financing, valuation, repositioning, or sale strategy.

A renovation decision is not complete when the contractor leaves the building. It is complete only when the hotel has assessed whether the project delivered the intended business result. This post-renovation discipline is essential because capital investment should create measurable operational, financial, or risk-related improvement.

Without measurement, the hotel may celebrate completion without knowing whether value has actually been created.

CHAPTER SUMMARY

Managing renovation programs requires a combination of engineering awareness, financial understanding, operational coordination, and executive judgment. A hotel renovation is never only about replacing equipment or refreshing the product. It is a decision about risk, cash, timing, guest impact, compliance, asset value, and strategic positioning.

The first lesson of this chapter is that renovation is often forced, not optional. Asset aging, brand standards, regulatory compliance, and market pressure can all create conditions where delay becomes more dangerous than action. The second lesson is that renovation must be managed as a lifecycle, from diagnostic and strategic decision to budgeting, phasing, execution, commissioning, and post-impact evaluation. Hotels that underestimate the first and last stages often suffer overruns and incomplete results.

The third lesson is that engineering and cosmetic renovation must be compared carefully. Cosmetic projects are visible and commercially attractive, but engineering projects often carry greater risk if delayed. The fourth lesson is that CAPEX and OPEX must be understood together. Delaying CAPEX may create OPEX traps through emergency repairs, energy waste, guest compensation, and staff firefighting. The fifth lesson is that hidden costs must be included in the business case. Room revenue loss, guest compensation, staff overtime, energy inefficiency, and management time can significantly change the real cost of renovation.

Finally, the chapter emphasizes that renovation arbitration requires mature leadership. The manager must prioritize compliance and safety, balance guest impact and financial survival, name the risks accepted, and communicate decisions in the language of owners and finance. The best renovation decisions are not necessarily the most visible or the cheapest. They are the decisions that protect the guest, the team, the asset, the brand, and the long-term value of the business.

OPERATIONAL CASE STUDY

Renovation Arbitration Under Budget Constraints

You are part of the Executive Committee of a four-star urban hotel. The property has 200 guest rooms, a full-service restaurant, several meeting rooms, and a loyal corporate client base. The hotel is independently operated but owned by an investment group that expects disciplined capital planning and improved asset performance over the next three years.

The hotel currently has a cash reserve of EUR 1.2 million available for renovation investment this year. Unfortunately, three urgent needs have emerged at the same time. The executive committee can fully fund only one project immediately. The other two projects must either be postponed, partially phased, or addressed through temporary OPEX measures.

Project	Investment Required	Current Situation
HVAC full replacement	EUR 900,000 CAPEX	System is at end of useful life; energy efficiency is declining; comfort complaints are increasing
Guest bathroom renovation	EUR 1,500,000 CAPEX	Bathrooms are outdated; online reviews mention weak product quality; renovation would support rate repositioning
Electrical compliance upgrade	EUR 400,000 CAPEX	Panels are not aligned with current compliance expectations; insurer has requested clarification after inspection

The owner wants the executive committee to defend a recommendation within 48 hours. The Finance Director is concerned about cash preservation. The Director of Sales argues that bathroom renovation would be the easiest project to explain commercially. The Chief Engineer warns that the HVAC system is becoming unreliable and that emergency repairs are increasing. The insurer has asked for documentation related to the electrical compliance issue. The General Manager must make a recommendation that is credible to ownership and responsible toward guests, staff, and the asset.

Executive Analysis

The first project to assess is the electrical compliance upgrade. Although it is the least expensive of the three, its strategic importance may be high because compliance and insurance exposure cannot be treated as optional. If the insurer is already asking questions, the hotel may face increased scrutiny, reduced coverage, or conditions attached to continued insurance support. The executive committee must determine whether this project is legally or operationally mandatory. If it is, it should be prioritized ahead of cosmetic improvement.

The second project is the HVAC replacement. The system affects guest comfort, energy cost, operational continuity, and future OPEX. If the system fails during high occupancy or extreme weather, the hotel may suffer room closures, compensation, complaints, and reputational harm. Even if the system does not fail immediately, inefficient performance may be eroding margin every month. The committee should request data: complaint frequency, energy cost trend, emergency repair history, number of rooms exposed, spare parts availability, and estimated cost of two additional years of patching.

The third project is the bathroom renovation. This project is highly visible and commercially attractive. It may support guest satisfaction and ADR growth. However, if the bathrooms are outdated but still functional, the project may be strategically important but less urgent than compliance or system failure risk. The committee must avoid emotional prioritization. Visible does not always mean critical.

Recommended Decision Logic

A defensible recommendation would likely prioritize the electrical compliance upgrade first if it is confirmed as mandatory or materially linked to insurance and legal exposure. Because the required amount is EUR 400,000, this still leaves EUR 800,000 of the cash reserve. The committee could then propose a phased HVAC strategy using the remaining funds, either beginning with the most exposed zones or funding preparatory works, equipment deposits, or the highest-risk components. Bathroom renovation could be postponed to the following budget cycle, with limited OPEX improvements focused on maintenance, cleaning quality, and targeted fixture replacement to reduce guest dissatisfaction temporarily.

Another possible recommendation would prioritize HVAC first if the electrical issue is judged manageable through immediate corrective compliance documentation and a short-term action plan. However, this option must clearly explain why compliance risk is acceptable. A weak recommendation would fund bathrooms first simply because they are visible, while leaving electrical compliance unresolved and HVAC failure risk uncontrolled.

The strongest executive briefing should not hide the trade-off. It should say clearly: "We recommend addressing compliance first because it is non-negotiable, then using remaining capital to reduce HVAC risk through phased replacement. We accept the commercial downside of postponing full bathroom renovation, but we do not accept unresolved compliance exposure or uncontrolled infrastructure failure risk."

Managerial Lessons from the Case

This case shows that renovation arbitration is rarely about choosing between a good project and a bad project. All three projects may be justified. The challenge is to rank them according to risk, timing, financial logic, and strategic consequence. Compliance projects must be treated with discipline. Engineering systems must be evaluated through lifecycle cost and operational exposure. Cosmetic projects must be connected to revenue logic but not allowed to override safety or asset integrity.

The case also shows that a limited budget does not excuse poor decision-making. When capital is constrained, the manager must become more precise, not less. The executive committee must quantify risks, identify hidden costs, build phasing options, and communicate accepted trade-offs transparently.

CHAPTER KEY TERMS

- **CAPEX:** Capital expenditure. A long-term investment in the hotel asset, usually capitalized and depreciated over time. In engineering renovation, examples include HVAC replacement, electrical infrastructure upgrade, boiler replacement, or complete plumbing renewal.

- **OPEX:** Operating expenditure. A recurring operating cost that affects the hotel's profit and loss performance more immediately. Maintenance contracts, emergency repairs, outsourcing, energy overconsumption, and temporary patches are typical engineering-related OPEX items.
- **Deferred Maintenance:** Necessary repair, replacement, or upgrade work that has been postponed. Deferred maintenance can temporarily protect cash but often increases future risk, operating cost, and probability of failure.
- **Lifecycle Cost:** The total cost of owning, operating, maintaining, repairing, and eventually replacing an asset over its useful life. Lifecycle cost is more useful than initial purchase cost when evaluating engineering renovation decisions.
- **Asset Integrity:** The condition in which a building system remains reliable, compliant, safe, and capable of performing according to required standards. Asset integrity protects both operational continuity and long-term value.
- **Phasing Strategy:** The planned sequencing of renovation work to manage cash flow, reduce operational disruption, align with seasonality, protect guest experience, and control execution risk.
- **Operational Continuity:** The hotel's ability to maintain service delivery during renovation or technical disruption. It requires coordination between Engineering, Rooms Division, Front Office, Sales, Revenue, Security, and Finance.
- **Guest Displacement:** The loss or relocation of room inventory caused by renovation works. Guest displacement

affects revenue, reservations, room assignment, group commitments, and guest satisfaction.

- **Hidden Costs:** Costs not always included in the initial project quote, such as room revenue loss, guest compensation, staff overtime, energy inefficiency, temporary systems, and internal management time.
- **Commissioning:** The process of testing and verifying that renovated or newly installed systems perform as intended before they are fully handed over to hotel operations.
- **Compliance Trigger:** A legal, regulatory, insurance, or safety requirement that makes a renovation mandatory rather than discretionary.
- **OPEX Trap:** A situation where repeated operating expenses are used to delay necessary capital investment, eventually producing higher total cost, greater disruption, and increased operational risk.

CHAPTER REVIEW QUESTIONS

- Why should hotel renovation be considered a strategic financial decision rather than only a technical project?
- Explain the difference between CAPEX and OPEX in the context of engineering renovation. Why can OPEX sometimes become "CAPEX in disguise"?
- What are the main drivers that force hotels to renovate, and how should a manager prioritize them?
- Why is it dangerous to prioritize cosmetic renovation over engineering renovation without a risk assessment?

- Describe the five main stages of a renovation lifecycle and explain why diagnostic work and post-renovation evaluation are often underestimated.
- What hidden costs should be included when preparing a renovation business case for ownership?
- How can poor phasing damage guest experience, staff workload, and financial performance?
- What questions should Finance ask before approving a major engineering renovation?
- In the operational case study, which project would you prioritize first and why? Explain the risks you accept by delaying the other projects.
- How should a General Manager communicate a renovation recommendation to owners without using technical jargon?

CHAPTER REFERENCES

- Course 5 presentation: Managing Renovation Programs: CAPEX & OPEX. Basics of Hospitality Engineering course materials.
- Course 5 teacher notes: Session 5 - Managing Renovation Programs: CAPEX & OPEX. Engineering Projects & Financial Logic.
- International Facility Management Association. Facility management body of knowledge and lifecycle management resources.
- ASHRAE. HVAC systems, energy efficiency, and building performance guidance for commercial buildings.

- RICS. Guidance on lifecycle costing, asset management, and capital project planning.
- International Energy Agency. Energy efficiency in buildings and operational performance resources.
- Cornell Hotel and Restaurant Administration Quarterly / Cornell Hospitality Reports. Research on hotel asset management, renovation, and operational performance.
- Hotel Asset Managers Association. Industry perspectives on hotel capital expenditure planning and asset value protection.

Chapter 6
Local Engineering Team Management

People, Leadership & Organization

Chapter General Overview

Engineering teams are often judged only when something breaks. Guests rarely notice the technician who prevented a boiler failure, the engineer who verified emergency lighting before a night audit, or the manager who negotiated a reliable subcontractor agreement before the next high-occupancy period. Yet these decisions shape the daily stability of a hotel. A hotel can have beautiful rooms, strong service culture, and ambitious commercial targets, but if its technical team is badly organized, under-skilled, poorly supervised, or exhausted by weak shift planning, the whole operation becomes fragile.

This chapter examines local engineering team management as a leadership discipline. It builds on the previous chapters by moving from systems to people. Earlier chapters established that engineering protects business continuity, guest comfort, life safety, communication quality, and asset value. This chapter asks a practical managerial question: who makes those outcomes possible at property level, and how should that team be structured, recruited, supervised, and evaluated?

The core message is simple: engineering is not only maintenance work. It is a management function. It requires clear reporting lines, realistic staffing, strong documentation discipline, cross-functional communication, carefully governed subcontractors, and fair but robust performance monitoring. The Engineering Manager is not merely the best technical person in the building. He or she must translate technical constraints into operational choices, lead technicians, coordinate vendors, protect compliance, and report risks in a way that the General Manager, owner, and asset manager can understand.

For future hospitality managers, this chapter is particularly important because team design reflects risk tolerance. A hotel that relies on a very small internal team and many subcontractors is making one kind of risk decision. A luxury hotel with specialist technicians and stronger redundancy is making another. Neither model is automatically right or wrong. The right structure depends on asset complexity, guest expectations, legal obligations, owner priorities, local labor market conditions, and the level of operational risk the hotel is willing to accept.

By the end of the chapter, students should be able to evaluate an engineering team from an executive perspective, distinguish technical competence from leadership capability, analyze the strengths and risks of internal teams and subcontractors, design shift and on-call logic, and draft a credible job description for a Hotel Engineering Manager.

Executive idea

Engineering teams do not usually create revenue directly. They protect revenue by keeping the hotel safe, functional, compliant, and ready to serve guests without interruption.

CHAPTER CONTENT

Engineering teams as protectors of revenue

Hotels operate twenty-four hours a day. Unlike many businesses, there is no real closing time during which the building can simply pause. Guests sleep, arrive, shower, connect, eat, use elevators, attend events, and expect comfort at every hour. In this environment, the engineering team protects the conditions that allow the commercial promise of the hotel to remain credible. The team may not sell rooms, but it keeps rooms sellable. It may not serve breakfast, but it keeps kitchen extraction, refrigeration, electricity, water, and safety systems reliable. It may not handle guest check-in, but it protects the elevators, door locks, lighting, and building systems that make the arrival experience possible.

This is why engineering team management must be understood as revenue protection. A poor staffing decision can become a guest complaint. A weak on-call system can become a refund. A poorly supervised subcontractor can become a safety exposure. A missing maintenance log can become an insurance problem. The financial link is sometimes indirect, but it is real. When technical issues accumulate, the property loses rooms to out-of-order status, increases compensation costs, weakens online reviews, and places greater pressure on front-line teams.

The managerial difficulty is that success is often invisible. When engineering is well managed, there may be fewer emergencies, fewer complaints, better energy performance, stronger compliance evidence, and a calmer operational rhythm. Because these outcomes are less visible than a crisis, owners and executives may underestimate the value of staffing, training, documentation, and preventive organization. A mature hotel leader must be able to defend invisible performance before failure makes it visible.

Engineering as a management function

A frequent mistake is to reduce engineering to repair work. This is too narrow. Repair is one visible part of the job, but engineering management covers a broader field: asset lifecycle management, CAPEX and OPEX discipline, compliance oversight, vendor governance, safety monitoring, preventive maintenance planning, energy control, and cross-functional communication.

The Engineering Manager therefore occupies a hybrid position. He or she must understand enough technical reality to make credible decisions, but must also manage people, schedules, reporting, documentation, budget priorities, and internal service relationships. A technician may focus on restoring an air-conditioning unit. The Engineering Manager must ask why the failure occurred, whether similar units are at risk, how many rooms could be affected, whether preventive maintenance is properly funded, and how the issue should be communicated to Operations and the General Manager.

In this sense, engineering management is closer to operational risk governance than to simple maintenance coordination. The manager must continuously connect technical facts to business implications. This connection is especially important in hotels with aging assets, limited capital expenditure, strong guest expectations, and complex ownership structures.

Local engineering and corporate engineering

Many hotel groups distinguish between local engineering at property level and corporate engineering at brand or ownership level. Property-level engineering is close to execution. It deals with immediate issues, local regulations, physical building conditions, and the daily reality of guests and staff. Corporate engineering, by contrast, often defines standards, approves major capital expenditure, protects brand consistency, and supports long-term asset strategy.

The relationship between these two levels can create productive tension. Corporate standards are useful because they prevent each property from improvising its own safety, maintenance, and investment logic. They help protect brand quality and ensure consistency across the portfolio. However, local reality matters. A building may be old, constrained by its historical structure, affected by local regulations, or facing urgent operational problems that a standard document cannot fully capture.

The General Manager and Engineering Manager must therefore mediate between standardization and local adaptation. They should not ignore corporate requirements, but they must communicate local risk clearly and factually. When corporate standards cannot be applied immediately, local leadership must document the constraint, propose a mitigation plan, and define a timeline. This is governance in action: not blind compliance, not local improvisation, but structured decision-making.

The managerial reality of engineering leadership

The local Engineering Manager usually leads technicians rather than engineers. In many properties, the team is composed of multi-skilled maintenance staff who deal with practical, everyday problems: guestroom defects, lighting issues, minor plumbing repairs, furniture repairs, equipment checks, and first response to system alarms. This requires technical breadth, flexibility, and service awareness. It also means that the manager must support people who are often under pressure, interrupted frequently, and expected to solve problems quickly with limited resources.

Engineering managers also work with imperfect assets. Many hotels are not new buildings. Systems may be old, documentation may be incomplete, previous renovations may have been partial, and equipment may have been maintained through years of patches. In such environments, the manager must maintain operational discipline without pretending that the asset is ideal. This requires honesty, prioritization, and strong reporting.

A final reality is that engineering is often judged when it fails, not when it succeeds. Guests, executives, and owners notice the broken elevator, the cold room, the leak, the delayed repair, or the fire system non-compliance. They rarely notice the hundreds of preventive actions that avoided similar issues. This creates emotional pressure on engineering teams and can damage morale. Good leadership therefore includes recognition, clarity, fair workload planning, and visible support from hotel management.

Typical engineering team structures

Engineering team structure varies significantly according to hotel size, positioning, age, asset complexity, service level, and ownership model. A small or mid-size hotel may operate with an Engineering Manager, one to three technicians, and a network of external contractors. This model can be efficient if the building is relatively simple, if contractor response times are reliable, and if preventive maintenance is well organized. However, the model can become risky when the hotel has complex systems, high occupancy, high guest expectations, or weak subcontractor availability.

A large or luxury hotel usually requires a more layered structure. It may include a Chief Engineer or Director of Engineering, shift engineers, technicians, and specialist roles for HVAC, electrical systems, plumbing, energy management, or life safety coordination. It will also rely on a strong subcontractor network for elevators, fire systems, kitchen equipment, BMS, boilers, chillers, and other specialized systems. In such properties, the engineering structure reflects the level of operational risk the hotel is unwilling to accept.

The key principle is that structure should follow risk, not tradition. A hotel should not ask, 'How many technicians did we have last year?' It should ask, 'What risks must this team control, what systems must it cover, what response time do we promise, and what level of guest disruption is acceptable?' Once these questions are answered, the staffing model can be evaluated more realistically.

Reporting lines and accountability

In many hotels, engineering reports directly to the General Manager. In other structures, it may report to the Operations Director, with indirect accountability to ownership or asset management for capital expenditure and asset protection. The reporting line matters because engineering decisions affect safety, compliance, guest satisfaction, and long-term property value. If engineering is buried too deeply in the organization, risks may not reach the right decision-makers in time.

Clear reporting lines also distinguish execution from accountability. The technician may execute a repair. The Engineering Manager owns the process, prioritization, and follow-up. The General Manager retains ultimate operational accountability for the property. If an air-conditioning complaint occurs at 2 a.m., the immediate response may involve the duty manager and technician, but the underlying responsibility includes staffing logic, preventive maintenance, guest recovery procedures, and escalation rules. These are management decisions.

Hotels must therefore avoid vague responsibility. If everyone assumes someone else owns the problem, delays occur. A strong engineering organization defines who can authorize urgent spending, who contacts subcontractors, who informs the duty manager, who updates the guest profile, who logs the incident, and who reports recurring issues at the next executive meeting.

Recruitment and skills mix

Recruiting for hotel engineering is not simply a matter of finding the most technically skilled person available. Technical capability is necessary, but insufficient. The department requires a balanced skills mix: multi-system competence, safety awareness, documentation discipline, operational communication, vendor coordination, and service culture. A hotel technician enters guest areas, communicates with non-technical colleagues, responds to urgent requests, and sometimes interacts directly with guests. Technical ability without communication maturity can create friction.

Safety awareness is particularly important. Engineering staff must understand that shortcuts can create legal, operational, and human risk. A poorly documented intervention, a blocked fire door, a delayed inspection, or a temporary repair left in place for too long can have consequences beyond the technical task itself. Documentation discipline is also essential. Logs, checklists, certificates, and maintenance records protect the hotel and prove that systems are being managed properly.

Operational communication is the fourth pillar. Engineering does not operate alone. It works daily with Front Office, Housekeeping, Food and Beverage, Security, Finance, and Sales. A successful engineering team knows how to explain constraints, coordinate room access, manage priorities, and translate technical issues into operational impact. This is why hiring only 'the best technician' can be a mistake if the role also requires leadership, communication, and judgment.

From technician to manager

Promoting the best technician to Engineering Manager may seem logical, but it can be risky if the promotion is treated as a reward rather than a leadership transition. A strong technician fixes problems. A strong manager prevents problems, plans work, coordinates people, explains priorities, manages budgets, and reports business risks. The two roles overlap, but they are not the same.

The shift from technician to manager requires training and support. The new manager must learn how to delegate, hold team members accountable, communicate with executives, negotiate with subcontractors, plan preventive maintenance, manage conflict, and prioritize under pressure. Without this support, the hotel may lose an excellent technician and gain an overwhelmed manager.

The right decision is not automatically to avoid internal promotion. Internal promotion can be powerful because the person knows the building, the team, the systems, and the operational culture. But the promotion should be accompanied by coaching, clear authority, performance expectations, and managerial development. Management is a learned skill, not a medal for technical excellence.

Recruitment challenges in hotel engineering

Many hotels face a shortage of skilled technicians. The labor market may favor construction, manufacturing, facilities management, or specialized contractors, leaving hotels competing for talent. Hospitality engineering can be perceived as less attractive because it involves irregular hours, guest pressure, emergency calls, and a wide range of tasks rather than one narrow specialty.

The aging workforce creates another challenge. Experienced technicians may hold valuable tacit knowledge about the building, but this knowledge is sometimes undocumented. When they leave, the hotel may lose practical understanding of hidden valves, old electrical panels, recurring defects, and informal repair history. This risk reinforces the need for documentation and knowledge transfer.

To attract and retain talent, hotels must present engineering as a professional career path, not merely a back-of-house repair function. This includes fair scheduling, safety culture, training opportunities, recognition, modern tools, and clarity about progression. HR and Engineering must work together: recruitment is not only a technical issue, it is an employer-brand issue.

Internal teams and subcontractors

The decision between internal staffing and subcontracting is one of the most important organizational choices in hotel engineering. Internal teams offer speed, institutional knowledge, daily availability, and better control. They understand the building, the guest rhythm, the back-of-house constraints, and the recurring issues that may not appear in formal reports. They can respond quickly to urgent problems and coordinate with other departments in real time.

Subcontractors offer specialized expertise, flexibility, and variable cost. They are essential for systems such as elevators, fire alarms, sprinklers, boilers, chillers, kitchen equipment, and certain electrical or regulatory inspections. A hotel should not pretend that its internal team can safely handle every complex system. In some areas, outsourcing is not only efficient, but necessary.

However, subcontracting creates dependency risk. If the external provider is slow, poorly supervised, or unclear about responsibilities, the hotel remains exposed. Outsourcing execution does not outsource accountability. The Engineering Manager must govern contractors through contracts, service-level expectations, documentation, response-time monitoring, compliance evidence, and performance review. The most realistic model for many hotels is hybrid: a small internal core team for daily operations, supported by external specialists for complex systems.

Shift organization and on-call logic

Shift organization expresses the hotel's risk tolerance. A property may choose full twenty-four-hour engineering coverage, limited coverage with night on-call, or a model where the duty manager coordinates first response after hours. Each option has financial and operational consequences. Lower payroll may look attractive, but if response times increase, guest complaints rise, minor defects worsen, and staff stress increases, the apparent saving may become expensive.

The decision depends on property size, occupancy pattern, system complexity, guest profile, brand promise, and legal constraints. A luxury hotel with high average daily rate and demanding guests may need stronger coverage than a limited-service property with simpler systems. A hotel with aging infrastructure may need more robust on-call logic than a new building. A property with large events, restaurants, or conference facilities must also consider peak operational risk outside normal office hours.

On-call arrangements are not free. They create burnout risk, motivation issues, potential compliance questions, and work-life balance concerns. Best practice requires a clear rotation schedule, fair compensation, escalation procedures, and realistic limits. If on-call becomes a hidden expectation rather than a structured arrangement, the hotel may damage morale and increase turnover.

Performance, reporting, and KPIs

Engineering performance must be measured in ways that connect to hotel outcomes. Counting work orders alone is insufficient because it says little about risk, guest impact, recurrence, or prevention. A team may close many work orders and still be operating reactively. A better performance framework includes guest-impacting incidents, preventive maintenance completion, repeat defects, response times for urgent issues, equipment downtime, energy performance, compliance documentation, subcontractor performance, and safety inspection results.

Good KPIs must be understood by management. They should answer practical questions: Are we preventing failures or reacting to them? Which systems create the most guest complaints? Are safety checks completed and documented? Are subcontractors meeting service levels? Are energy costs moving in the right direction? Which assets require capital planning?

Reporting should be concise and decision-oriented. The Engineering Manager should not present only technical lists. Reports should identify risk level, business impact, required decisions, budget implications, and timeline. A monthly engineering report should help the General Manager decide, not simply observe.

The Hotel Engineering Manager job description
A job description for a Hotel Engineering Manager should not read like a list of repairs. It should describe a leadership role. The purpose of the role is to ensure technical reliability, guest comfort, life safety, asset protection, compliance, cost control, and operational continuity. The job description must clarify reporting lines, decision

rights, team management responsibilities, technical scope, vendor governance, documentation obligations, and key performance indicators.

A strong job description includes authority. Can the manager approve emergency spending up to a defined limit? Can he or she call subcontractors without prior approval during life-safety or guest-impacting incidents? What decisions require General Manager or owner approval? Without decision rights, accountability becomes unfair. The hotel cannot demand results from a manager who lacks the authority to act.

The job description should also define success. Example indicators may include preventive maintenance completion rate, reduction in repeat guestroom defects, compliance documentation accuracy, response time to critical incidents, energy consumption performance, room out-of-order reduction, subcontractor response compliance, and team training completion. These measures connect the role to business value rather than only technical activity.

Management routines and daily discipline

Strong engineering teams are not built only through organizational charts. They are built through daily and weekly routines that make risk visible before it becomes a guest incident. A short morning briefing can align the team on urgent rooms, VIP arrivals, out-of-order rooms, safety checks, contractor visits, and unresolved issues from the night. A weekly meeting with Rooms Division can identify repeat defects, priority floors, guestroom patterns, and housekeeping observations. A monthly executive review can connect engineering work to energy cost, compliance status, asset concerns, and capital expenditure priorities.

These routines create discipline. Without them, engineering becomes reactive and fragmented. Technicians respond to the loudest complaint, Front Office becomes frustrated by a lack of information, Housekeeping blocks rooms without visibility, and the General Manager hears about problems only when they become complaints. Routines do not eliminate technical failures, but they reduce surprise and improve coordination.

A practical routine should be simple enough to survive operational pressure. It should define what is reviewed, who attends, what decisions are expected, and how follow-up is tracked. The Engineering Manager should leave each meeting with a short list of decisions, priorities, and owners. Good management is not only what is known; it is what is followed up.

Cross-functional coordination with operations

Engineering performance is inseparable from cross-functional coordination. Front Office needs realistic repair times so it can communicate with guests, manage room moves, and avoid promising what cannot be delivered. Housekeeping needs clear status updates so rooms are not released too early or blocked unnecessarily. Food and Beverage needs preventive coordination around kitchen equipment, refrigeration, extraction, dishwashing, lighting, and event setup. Security needs alignment on fire systems, emergency lighting, access control, and incident response.

Many engineering conflicts are not caused by bad will. They are caused by timing, language, and unclear priorities. A technician may believe that a repair is minor because the equipment can still function. Front Office may see the same issue as urgent because the guest is angry. Housekeeping may see it as operationally disruptive because a room cannot be cleaned, inspected, or sold. The

Engineering Manager must help departments understand the difference between technical severity and guest severity. Both matter, but they are not always the same.

The best hotels create shared priority rules. For example, life-safety issues come first, then guest-occupied rooms, then rooms needed for same-day arrivals, then recurring defects, then aesthetic repairs. These rules prevent every department from treating its own request as the only priority. They also protect the engineering team from chaotic demand.

Knowledge transfer and documentation culture

Engineering departments often depend on informal knowledge. A senior technician knows which valve controls a particular riser, which guestroom has a recurring condensation issue, which contractor responds fastest, and which panel was modified during a previous renovation. This knowledge is valuable, but it is also dangerous if it remains only in one person's memory. When that person leaves, retires, or is absent, the hotel loses operational intelligence.

Documentation culture is therefore a leadership priority. It includes maintenance logs, asset registers, warranties, contractor reports, commissioning records, inspection certificates, equipment manuals, spare-parts lists, emergency procedures, and simple building maps that can be used by new staff. The purpose is not bureaucracy. The purpose is continuity. A hotel that cannot locate its own technical information is exposed during emergencies, audits, insurance claims, and staff turnover.

The Engineering Manager should gradually convert tacit knowledge into shared knowledge. This can be done through checklists, photographs, annotated plans, digital folders, and short internal briefings. The goal is not to document everything perfectly at once, but to build a habit: every important intervention should leave a trace that helps the next person understand the system better.

Coaching, motivation, and team culture

Engineering work can be physically demanding and psychologically frustrating. Technicians are often called when someone is unhappy, when something has failed, or when a department is under pressure. Their successes are preventive and invisible; their failures are public and immediate. This makes motivation and recognition essential. A team that feels ignored may become purely reactive, defensive, or disengaged.

The Engineering Manager must create a culture where technicians understand the hospitality meaning of their work. A repaired shower is not only a technical task; it protects a guest's morning. A tested emergency light is not only compliance; it protects life safety. A completed preventive inspection is not only a checklist; it prevents future disruption. When technicians understand the service value of their work, they are more likely to communicate, document, and prioritize with care.

Coaching also means helping technicians develop beyond immediate repair skills. A strong technician should learn how to speak to colleagues, report clearly, read priorities, use the maintenance management system, interact respectfully in guest areas, and understand the financial consequences of repeat defects. These capabilities increase both team performance and career attractiveness.

General Manager questions
A GM should regularly ask: Do we have the right skills for our risk profile? Are we too dependent on one person or one contractor? Are preventive tasks really completed? Are night incidents covered

fairly and effectively? Do our engineering reports lead to decisions?

CHAPTER SUMMARY

Local engineering team management is about organizing people to protect revenue, safety, compliance, asset value, and guest experience. The engineering team may be invisible when it succeeds, but its absence or weakness becomes immediately visible when a system fails. Managing this team therefore requires executive discipline, not only technical knowledge.

The chapter has emphasized that structure reflects risk tolerance. Small hotels may rely on lean internal teams and external contractors, while large or luxury properties often require specialized roles and stronger redundancy. Neither model is automatically correct. What matters is whether the structure matches the property's operational complexity, guest expectations, asset condition, and legal obligations.

Recruitment must look beyond technical skills. A hotel engineering team needs safety awareness, documentation discipline, communication ability, and service understanding. Promoting the best technician can be successful only if the hotel supports the transition from technical execution to leadership performance. Management is a learned skill, not a reward.

Subcontractors are essential, but they require governance. Outsourcing technical execution does not outsource accountability. The Engineering Manager must coordinate vendors, monitor service levels, preserve documentation, and ensure that external expertise supports the hotel's risk profile. Similarly, shift and on-call decisions must balance payroll control against guest satisfaction, staff wellbeing, and asset protection.

The final managerial lesson is that the Engineering Manager role must be clearly defined. Strong job descriptions include role purpose, decision rights, reporting lines, technical scope, team management, vendor governance, and business-oriented KPIs. People decisions protect assets. The right engineering team safeguards the hotel's ability to operate.

OPERATIONAL CASE STUDY

The Riverside Meridian Hotel

The Riverside Meridian is a 180-room, four-star urban hotel located near a business district and leisure waterfront area. The property serves weekday corporate travelers, weekend leisure guests, small conferences, and private events. It has a full-service restaurant, three meeting rooms, a small fitness area, and an underground parking facility. The building is twenty-two years old. Some infrastructure has been updated, but the hotel still operates with aging HVAC components, recurring plumbing issues, and several systems that require external contractors.

The internal engineering team consists of one Engineering Manager, two technicians, and a part-time administrative assistant shared with Housekeeping. Elevators, fire systems, kitchen extraction, and major HVAC works are outsourced. The Engineering Manager was promoted two years ago after being recognized as the strongest technician on the team. He knows the building extremely well, but he has received no formal management training. He is respected for his technical competence but struggles with delegation, documentation, and communication with the Executive Committee.

Over the last six months, guest complaints have increased. Most relate to air-conditioning noise, slow response to maintenance requests, and water temperature inconsistency during morning peak periods. Front Office complains that Engineering does not always update repair status in the system. Housekeeping says rooms remain blocked longer than necessary because technicians do not always confirm when a repair is complete. Finance is concerned that subcontractor emergency invoices are rising. The owner has asked the General Manager to reduce payroll and control maintenance cost.

At the same time, the team is showing signs of fatigue. One technician regularly covers evening calls after finishing his daytime shift. The other technician has expressed frustration that subcontractors are called too late, forcing the internal team to improvise temporary fixes. The Engineering Manager feels trapped between guest expectations, limited CAPEX, owner pressure, and staff fatigue. He insists that he can keep the building running, but his reports to the GM are mostly technical and do not clearly identify risk, urgency, or decisions required.

A critical incident occurs on a Tuesday night at 2:15 a.m. The hotel is 94 percent occupied. A corporate group is arriving early the next morning, and several VIP rooms are allocated on upper floors. A guest calls Front Office to report that the room temperature is too high and that the air-conditioning unit is making a loud vibration. Within thirty minutes, four additional rooms on the same floor report similar issues. The night duty manager calls the on-call technician, who takes forty-five minutes to arrive. He identifies a likely HVAC distribution issue but cannot solve it without a specialist contractor. The contractor cannot come before 8:30 a.m.

By morning, twelve guests have complained, three rooms have required compensation, two guests have posted negative comments online, and the Front Office Manager is angry that there was no clear escalation process. The General Manager now asks for a review of the engineering team structure, on-call logic, subcontractor governance, and Engineering Manager job description.

Case analysis questions

- What is the difference between execution and accountability in this incident?
- Which problems are technical, and which are organizational or managerial?
- Was the hotel's internal team structure appropriate for its operational risk profile? Why or why not?
- What should the on-call policy include to avoid similar delays and staff burnout?
- Which subcontractor governance mechanisms are missing?
- How should the Engineering Manager report this incident to the General Manager?
- What KPIs should the hotel introduce to monitor engineering performance?
- Should the best technician automatically remain Engineering Manager? What support might be required?
- What decision rights should be clarified in the Engineering Manager job description?
- What immediate and longer-term actions should the Executive Committee take?

Suggested executive response

The Executive Committee should avoid treating the incident as a single HVAC failure. The technical issue is only the visible symptom. The deeper problem is an engineering management system that lacks sufficient clarity around team structure, escalation, contractor response, on-call expectations, and reporting. The first action should be a structured post-incident review that identifies guest impact, response timeline, decision points, and documentation gaps.

Within forty-eight hours, the hotel should clarify escalation procedures for night incidents, confirm which systems require immediate contractor involvement, and define who can authorize emergency call-outs. Within thirty days, the GM should review the engineering staffing model, on-call rotation, subcontractor contracts, preventive maintenance plan, and monthly reporting format. The Engineering Manager may remain in position, but only if supported through management coaching, clearer decision rights, and measurable KPIs.

The owner should be briefed in business language: guest impact, compensation cost, online reputation risk, staffing pressure, subcontractor dependency, and potential CAPEX implications. The message should not be, 'The HVAC failed.' It should be, 'The property’s current engineering organization is not fully aligned with its risk profile, guest expectations, and asset condition.'

CHAPTER KEY TERMS

- **Accountability:** The ownership of outcomes, even when tasks are executed by someone else. In engineering management, accountability often remains with the Engineering Manager and General Manager, even when a technician or subcontractor performs the work.
- **Execution:** The practical completion of a task, such as repairing equipment, checking a system, or calling a contractor. Execution can be delegated; accountability cannot.

- **Engineering Manager:** The property-level leader responsible for organizing engineering work, supervising technicians, coordinating vendors, protecting compliance, and translating technical issues into operational and business priorities.
- **Chief Engineer:** A senior engineering leader, often found in larger or luxury properties, responsible for technical reliability, team organization, preventive maintenance, compliance, and strategic coordination with management.
- **Skills Mix:** The combination of technical, safety, communication, documentation, and leadership capabilities required within the engineering team.
- **Technical Expertise:** Specialized knowledge required to understand, diagnose, maintain, or repair building systems. It is essential, but not sufficient for leadership roles.
- **Leadership Capability:** The ability to plan, delegate, communicate, prioritize, coach, hold people accountable, and make decisions under pressure.
- **Internal Team:** Engineering staff employed directly by the hotel, usually responsible for daily operations, first response, preventive checks, and coordination with other departments.
- **Subcontractor:** An external specialist or company contracted to provide expertise or services that the internal team cannot or should not perform alone, such as elevator maintenance or fire system certification.
- **Hybrid Model:** An engineering organization combining a small internal core team with external specialists for complex or regulated systems.

- **Vendor Governance:** The structured supervision of external providers through contracts, response expectations, documentation, compliance evidence, and performance review.
- **On-Call Duty:** An arrangement requiring staff or contractors to be available outside normal working hours for urgent technical issues.
- **Shift Coverage:** The staffing model that determines whether engineering support is available during specific periods, including evenings, nights, weekends, and high-risk operational windows.
- **Preventive Maintenance Completion Rate:** A KPI measuring whether planned maintenance tasks are completed on schedule, helping assess whether the team is preventing failures rather than only reacting to them.
- **Repeat Defect:** A recurring technical issue in the same room, area, or system, often indicating that the root cause has not been solved.
- **Response Time:** The time between a reported issue and the start of engineering intervention. For guest-impacting incidents, this can strongly influence satisfaction and service recovery.
- **Decision Rights:** The authority given to a manager to approve action, spending, escalation, or vendor intervention without waiting for additional approval.
- **Reporting Line:** The formal chain through which the Engineering Manager reports performance, risk, budget needs, and incidents to senior management.

- **Asset Protection:** The organized effort to preserve building value, system reliability, equipment lifespan, and long-term owner investment.
- **Risk Tolerance:** The level of operational, financial, safety, or reputational risk a hotel is willing to accept through its staffing, maintenance, and investment decisions.

CHAPTER REVIEW QUESTIONS

- Why should local engineering team management be considered a leadership function rather than only a maintenance function?
- Explain how engineering teams protect hotel revenue even if they do not generate revenue directly.
- What is the difference between accountability and execution in a hotel engineering context?
- Why can promoting the best technician to Engineering Manager become a strategic risk if the promotion is not supported properly?
- Compare the advantages and risks of an internal engineering team and external subcontractors.
- Why is the hybrid model common in hotel engineering, and what does it require from the Engineering Manager?
- How does shift organization reveal a hotel's risk tolerance?
- Why is on-call duty not truly free from a managerial perspective?
- Identify five KPIs that would help a General Manager evaluate engineering team performance.

- What should a strong Hotel Engineering Manager job description include beyond technical duties?

CHAPTER REFERENCES

- American Hotel & Lodging Educational Institute. Hospitality Facilities Management and Design. Lansing, MI: AHLEI.
- Barrett, P., and Baldry, D. Facilities Management: Towards Best Practice. Oxford: Blackwell Science.
- Cotts, D. G., Roper, K. O., and Payant, R. P. The Facility Management Handbook. New York: AMACOM.
- International Facility Management Association. Operations and Maintenance Benchmarks and Best Practices. Houston: IFMA.
- Kandampully, J., Zhang, T., and Bilgihan, A. Customer loyalty: a review and future directions with a special focus on the hospitality industry. International Journal of Contemporary Hospitality Management.
- Mullins, L. J. Management and Organisational Behaviour. Harlow: Pearson.
- O'Fallon, M. J., and Rutherford, D. G. Hotel Management and Operations. Hoboken, NJ: Wiley.
- Ransley, J., and Ingram, H. Developing Hospitality Properties and Facilities. Oxford: Butterworth-Heinemann.
- Roper, K. O., and Payant, R. P. The Facility Management Handbook. New York: AMACOM.
- Stipanuk, D. M. Hospitality Facilities Management and Design. Lansing, MI: Educational Institute of the American Hotel & Lodging Association.

Conclusion and Appendices

Basics of Hospitality Engineering - A Practical Guide for Future Hotel Managers

CONCLUSION - ENGINEERING AS A MANAGERIAL DISCIPLINE

Engineering in hospitality is often described as a technical function because it deals with equipment, systems, infrastructure and buildings. Yet the deeper purpose of this book has been to show that engineering is also a managerial discipline. It is a way of protecting service continuity, guest comfort, safety, asset value, financial discipline and brand credibility. Future hotel managers do not need to become engineers. They do, however, need to become credible decision-makers when engineering issues affect the business.

A hotel is a living operational system. It depends on air, water, power, heat, light, safety systems, communication networks, elevators, kitchens, drainage, waste flows and many other invisible processes that must work at the same time. Guests rarely see these systems. They judge the result: a comfortable room, a safe corridor, a quiet night, a hot shower, a functioning elevator, a clean public area, and a sense that the hotel is under control. When engineering works well, it disappears. When it fails, it becomes the center of the guest experience.

Managerial principle
Engineering is not only about repairing what is broken. It is about preventing operational failure before it becomes visible to guests, staff, owners, regulators, insurers or the public.

From Technical Maintenance to Business Continuity

The first managerial shift is to understand engineering as business continuity. The Engineering Department protects the ability of the hotel to remain open, sell rooms, operate restaurants, receive events, maintain safety and honour the brand promise. A technical failure is rarely only technical. A failed air-conditioning system becomes a guest satisfaction issue. A water pressure problem becomes a service recovery issue. A fire alarm malfunction becomes a legal and reputational issue. A poorly planned renovation becomes a cash flow, scheduling and asset-value issue.

The hotel manager must therefore connect engineering realities to business consequences. The key question is not only: What equipment has failed? The stronger managerial questions are: Who is affected? What is the risk of doing nothing? What is the cost of delay? What is the impact on guest experience? What documentation exists? What decision is required?

Accountability Cannot Be Outsourced

Several chapters have repeated a central idea: execution can be delegated, but accountability cannot. A technician may repair the system. A contractor may inspect the fire alarm. A specialist may maintain the elevator. But the hotel remains responsible for safety, compliance, planning, budget decisions and operational readiness. The General Manager, the Executive Committee and the property leadership team must understand enough to challenge reports, ask for evidence, approve preventive action and ensure that risks are being managed.

This is especially true in fire, life safety and regulatory matters. The phrase I did not know is not a strong managerial defense. A serious hotel leadership culture requires documentation, training records, maintenance logs, inspection certificates, escalation procedures and clear responsibility lines. Good engineering management is therefore not only a matter of tools and skills. It is also a matter of governance.

Financial Discipline and Asset Protection

Engineering also belongs to the financial discipline of the hotel. Poor maintenance increases emergency costs. Deferred CAPEX often returns as uncontrolled OPEX. Energy inefficiency silently damages margins. Renovations that are badly phased reduce revenue and guest satisfaction at the same time. From an ownership perspective, engineering decisions protect the long-term value of the asset. From an operations perspective, they protect the ability to deliver service every day. From a finance perspective, they influence cash flow, EBITDA, depreciation, lifecycle cost and return on investment.

This is why the distinction between CAPEX and OPEX is more than accounting vocabulary. It shapes how managers think about time, risk and value. A cheap decision today may become an expensive problem tomorrow. A preventive investment may look costly at first but protect the property from disruption, negative reviews, regulatory exposure and emergency premiums.

People, Language and Leadership

Engineering is also a people function. The best technical strategy will fail if the team is poorly structured, undertrained, exhausted, badly supervised or unclear about priorities. A hotel needs the right balance between internal capability and external specialist support. It needs clear reporting lines, realistic on-call arrangements, vendor governance, documentation discipline and communication routines with Front Office, Housekeeping, F&B, Security, Finance and ownership.

Language is part of leadership. Future managers must learn to translate engineering issues into business impact. A technical report without risk level, urgency, cost of inaction and recommendation is incomplete from a managerial point of view. Strong managers do not hide behind jargon. They make engineering understandable enough for decisions to be taken.

> **Final executive insight**
> Hospitality engineering is the discipline of protecting the invisible conditions that make service possible. It is technical in its tools, financial in its consequences, human in its execution, and managerial in its accountability.

What Future Hotel Managers Should Remember

- Engineering is the invisible foundation of guest experience and business continuity.
- Technical failures quickly become service, financial, legal and reputational failures.
- Preventive maintenance is not a technical luxury; it is a management responsibility.
- CAPEX and OPEX decisions must be understood through lifecycle cost and risk exposure.
- Fire and life safety are leadership obligations, not only engineering procedures.
- The quality of reporting determines the quality of executive decisions.
- Engineering teams need leadership, structure, communication and governance.
- The role of the hotel manager is not to fix the system, but to understand the risk, ask the right questions and decide responsibly.

APPENDICE A : ENGINEERING MANAGER JOB DESCRIPTION TEMPLATE

This template supports the recruitment, evaluation or redesign of a Hotel Engineering Manager role. It is especially useful for a small or mid-size property where the Engineering Manager must combine technical coordination, team leadership, contractor governance, safety discipline and executive reporting.

Section	Template content
Job title	Hotel Engineering Manager / Chief Engineer
Reporting line	Reports to the General Manager or Operations Director. Works closely with Rooms Division, Housekeeping, F&B, Security, Finance, Human Resources and ownership or asset management representatives.
Role purpose	To ensure the safe, reliable and efficient operation of the hotel building, technical systems and engineering team. The role protects guest comfort, operational continuity, legal compliance, energy performance and asset value.
Core responsibilities	Maintain building systems and infrastructure. Coordinate preventive and corrective maintenance. Supervise technicians and subcontractors. Monitor fire and life safety systems. Manage documentation and compliance records. Support CAPEX planning and renovation works. Communicate technical risks in business language.
Technical scope	HVAC, plumbing, electrical distribution, emergency power, elevators, fire detection

	and suppression, kitchen equipment coordination, guestrooms, public areas, back-of-house, plant rooms, waste management and energy monitoring.
Team management	Plan shifts and on-call coverage. Allocate work orders. Train and coach team members. Ensure safety procedures are followed. Monitor workload, morale and productivity. Build collaboration with operations departments.
Subcontractor management	Select, brief and supervise external specialists. Validate scope of work, response times, documentation, certifications and service quality. Maintain service level agreements and evaluate vendor performance.
Budget and financial responsibility	Prepare maintenance budgets. Identify CAPEX needs. Control emergency call-out costs. Track energy consumption and maintenance expenses. Translate technical needs into financial priorities.
Compliance and safety	Maintain inspection records, statutory certificates, fire and life safety logs, risk assessments, permits, training records and incident documentation. Escalate all safety concerns immediately.
Key performance indicators	Guest-impacting incidents, preventive vs corrective maintenance ratio, equipment downtime, energy cost per occupied room, work order response time, compliance completion rate, out-of-order room days, subcontractor performance.

Required profile	Strong technical background, hotel or building maintenance experience, safety mindset, documentation discipline, leadership ability, clear communication skills, budget awareness and capacity to work under pressure.
Leadership expectations	The Engineering Manager is not only a technician. The role requires anticipation, prioritization, accountability, cross-functional communication and the ability to brief executives clearly.

APPENDICE B : EXECUTIVE INCIDENT BRIEFING TEMPLATE

This template helps managers translate a technical incident into a concise executive briefing. It should be used for incidents that affect guest comfort, safety, operations, compliance, revenue, reputation or asset protection.

Briefing element	Guidance
Incident title	Use a clear title that states the business issue, not only the technical component. Example: HVAC failure affecting guest comfort on floors 4 and 5.
Context	What happened? When did it happen? Where did it happen? Which systems, areas or departments are involved? Avoid excessive jargon.
Current status	Is the issue contained, ongoing or escalating? What has already been done? Who is currently managing the response?
Guest impact	How many guests, rooms, events or public areas are affected? Are relocations, compensation or service recovery actions required?
Operational impact	Which departments are affected? Are rooms out of order? Are services delayed? Is staffing under pressure?
Safety and compliance impact	Is there a life safety, legal, insurance or regulatory concern? Is documentation

	required? Has the relevant authority or insurer been informed if necessary?
Financial exposure	Estimate the cost of immediate response, revenue loss, compensation, energy impact, emergency contractor cost or potential CAPEX requirement.
Risk if no action is taken	Describe what could happen if the decision is delayed: guest complaints, system failure, safety exposure, revenue loss, negative reviews, inspection issue or asset damage.
Options	Present a maximum of two realistic options. Each option should include cost, timing, risk and guest impact.
Recommendation	State the recommended option clearly and explain why it is the best balance between safety, guest experience, cost and operational continuity.
Decision required	Specify exactly what approval or decision is needed, from whom, and by when.
Follow-up and documentation	Indicate the next reporting time, records to be kept, responsible person, and communication plan for guests, staff, owner or insurer.

Executive briefing example

Context: The main chiller serving floors 5 to 8 has lost capacity during a high-occupancy period. Guest room temperatures are rising above comfort levels and complaints have started at Front Office.

Risk: If no decision is taken within the next two hours, up to 72 rooms may require relocation or compensation. There is also risk of negative online reviews and additional equipment damage if the system is forced to run beyond safe operating limits.

Options: Option one is to approve emergency specialist intervention tonight at premium cost. Option two is to operate at reduced capacity until morning, with likely guest compensation and room moves. Recommendation: approve emergency intervention tonight because the guest and reputation exposure is greater than the premium call-out cost.

APPENDICE C :PREVENTIVE MAINTENANCE CHECKLIST

This checklist is a managerial tool. It does not replace manufacturer requirements, local regulations, statutory inspections or brand standards. It helps hotel leaders verify that preventive maintenance is planned, documented and connected to guest experience, safety and asset protection.

Daily and shift checks

Review overnight incidents, guest complaints and unresolved engineering tickets.
Check critical alarms, BMS dashboard or plant room indicators.
Verify hot water availability, pressure complaints and HVAC comfort issues.
Inspect main public areas for lighting, leaks, odors, trip hazards and visible defects.
Confirm elevator operation and report irregularities immediately.
Review out-of-order rooms and prioritize guest-impacting repairs.
Log all incidents, actions and pending follow-ups.

Weekly checks

Inspect guestroom sample set for HVAC function, lighting, plumbing, drainage and minor defects.
Check fire doors for obstruction, damage and correct closing behavior.
Review work order backlog and identify repeated issues by area or system.
Inspect back-of-house areas, kitchens, laundry, waste storage and staff areas.
Review subcontractor visits and verify documentation received.
Check emergency supplies, tools and critical spare parts.

Monthly checks

Review preventive maintenance completion rate and corrective maintenance ratio.
Test emergency lighting according to applicable requirements.

Review energy consumption trends and investigate abnormal movements.
Inspect HVAC filters, ventilation performance and comfort complaint trends.
Review water temperature, water pressure and water treatment records where applicable.
Review fire alarm test records, sprinkler inspection status and safety system documentation.
Meet with Front Office, Housekeeping and F&B to review recurring technical pain points.

Quarterly checks

Review asset condition of major systems: HVAC, boilers, chillers, pumps, electrical panels, elevators and plumbing.
Update engineering risk register and escalate high-risk items to GM or ownership.
Review subcontractor performance, service level agreements and response times.
Assess CAPEX needs and deferred maintenance exposure.
Conduct a cross-functional safety and operational readiness review.
Validate training needs for engineering and duty management teams.

Annual checks

Update the annual preventive maintenance plan and budget.
Review statutory inspection calendar, certificates and insurance requirements.
Prepare CAPEX plan using asset condition, guest impact, compliance risk and lifecycle cost.
Review emergency response procedures and conduct relevant drills.
Evaluate engineering team structure, skills mix, on-call policy and subcontractor strategy.
Review sustainability and energy performance targets for the next year.

APPENDICE D : CAPEX/OPEX DECISION GRID

This grid helps hotel managers arbitrate between immediate repair, ongoing OPEX patching, phased investment and full CAPEX replacement. It is designed for executive discussion and should be used with engineering, finance, operations and ownership.

Decision factor	**Managerial question**
Guest impact	Does the issue directly affect comfort, safety, access, sleep quality, water, temperature, noise, lighting or service continuity?
Revenue exposure	Does it create out-of-order rooms, event cancellations, compensation, lower rates, booking loss or negative reviews?
Safety and compliance	Is the issue linked to fire, life safety, accessibility, building code, environmental regulation, permits or insurance obligations?
Asset condition	Is the system near end of useful life? Are failures becoming frequent? Are spare parts difficult to obtain?
Lifecycle cost	Is the total cost of repeated repair, energy waste and downtime greater than replacement over the expected life of the asset?
Energy impact	Does the current system create abnormal utility cost, poor efficiency or sustainability reporting risk?

Operational disruption	Can the work be phased? What is the impact on rooms, restaurants, meetings, staff workload and guest perception?
Cash flow and timing	Can the investment be aligned with low season, ownership funding, financing cycles or planned closure periods?
Risk of deferral	What is likely to happen if the decision is delayed by 3 months, 12 months or 24 months?
Strategic importance	Does the investment support brand positioning, asset value, compliance, sustainability or long-term owner objectives?

Suggested scoring logic

Score each factor from 1 to 5. A score of 1 indicates low urgency or low impact. A score of 5 indicates high urgency, high impact or non-negotiable exposure. Any item involving life safety, legal compliance or imminent catastrophic failure should be treated as a priority regardless of total score.

Score	**Meaning**	**Typical action**	**Executive interpretation**
1	Low impact	Monitor	No immediate investment required
2	Moderate impact	Plan minor OPEX	Manage through routine maintenance

3	Growing risk	Prepare phased plan	Include in budget cycle
4	High risk	Prioritize CAPEX or urgent project	Escalate to EXCOM/owner
5	Critical or mandatory	Immediate action	Compliance, safety or business continuity priority

Decision outcomes

- Continue OPEX maintenance when risk is low, system reliability is acceptable and guest impact is limited.
- Use phased CAPEX when the system is aging but the hotel can sequence investment without unacceptable risk.
- Approve full CAPEX replacement when lifecycle cost, guest impact, compliance exposure or failure risk justify immediate investment.
- Reject cosmetic priority when mandatory safety, compliance or business continuity issues remain unresolved.
- Document all accepted risks when a project is postponed, including expected cost, guest impact and accountability owner.